Deep Learning for Numerical Applications with SAS®

Henry Bequet

Preface by
Oliver Schabenberger, Ph.D.
Executive Vice President, Chief Operating
Officer, and Chief Technology Officer, SAS

§.sas®

sas.com/books

The correct bibliographic citation for this manual is as follows: Bequet, Henry G. 2018. *Deep Learning for Numerical Applications with SAS®*. Cary, NC: SAS Institute Inc.

Deep Learning for Numerical Applications with SAS®

Copyright © 2018, SAS Institute Inc., Cary, NC, USA

ISBN 978-1-63526-680-1 (Hardcopy)
ISBN 978-1-63526-677-1 (EPUB)
ISBN 978-1-63526-678-8 (MOBI)
ISBN 978-1-63526-679-5 (PDF)

SAS Institute Inc., SAS Campus Drive, Cary, North Carolina 27513-2414.

July 2018

SAS® and all other SAS Institute Inc. product or service names are registered trademarks or trademarks of SAS Institute Inc. in the USA and other countries. ® indicates USA registration.

Other brand and product names are trademarks of their respective companies.

Contents

Preface

Artificial Intelligence (AI) and Machine Learning (ML) are all the rage. Computerized systems that can perform human tasks and make decisions are affecting many industries.

A core technology of these systems is deep learning, which is based on deep neural networks. Neural networks are not new, yet the successes in artificial intelligence are relatively recent. The availability of more computing power through multicore CPUs and Graphics Processing Units (GPUs) enabled us to train deeper networks. The availability of big data enabled us to train these networks well. The availability of specific neural networks—such as convolutional and recurrent networks—fueled the advances in image processing and natural language processing.

Combined, these forces created the perfect substrate for AI applications to grow.

Henry Bequet reminds us in this book that neural networks are algorithms to predict outcomes, to classify observations, and to detect patterns. They have many applications outside of computer vision, chatbots, and autonomous vehicles. The forces that accelerated progress through deep learning in cognitive analytics can be brought to bear in other domains, such as regression, function approximation, and Monte Carlo simulation.

In this book, Henry takes you on a tour of deep learning with SAS® using surprising applications that broaden your understanding of the technology. Henry guides you through the deep learning capabilities of SAS® Viya® that extend and complement your SAS experience.

Oliver Schabenberger, PhD
Executive Vice President, Chief Operating Officer and Chief Technology Officer
SAS

About This Book

What Does This Book Cover?

Machine learning and deep learning are ubiquitous in our homes and workplaces, from machine translation, to image recognition, to predictive analytics, to autonomous driving. Deep learning holds the promise of improving many of the applications that we use every day in a variety of fields. Most of the deep learning literature that is currently available explains the mechanics of deep learning with the goal of implementing cognitive applications fueled by big data. This book is different. Written by an expert in high-performance analytics, this book introduces a new field: deep learning for numerical applications (DL4NA). In contrast to deep learning, the primary goal of DL4NA is not to learn from data. The primary goal of DL4NA is to dramatically improve the performance of numerical applications by training deep neural networks.

This book presents the concepts and techniques step by step in a practical way so that you can easily reproduce the examples on your high-performance analytics systems. This book also discusses the latest hardware innovations that can power your SAS programs, including many-core CPUs, graphics processing units (GPU), field-programmable gate arrays (FPGA), and application-specific integrated circuits (ASIC).

Is This Book for You?

This book assumes no prior knowledge of high-performance computing, machine learning, or deep learning. It is for SAS developers and programmers who want to develop and run the fastest analytics.

It is also for those who are curious about the roots of deep learning and want an introduction to this fascinating field.

What Are the Prerequisites for This Book?

The prerequisites of this book are familiarity with SAS and the SAS programming language.

What Should You Know about the Examples?

This book includes tutorials for you to follow to gain hands-on experience with SAS.

Software Used to Develop the Book's Content

SAS 9.4 M5 (including SAS Studio)

SAS Viya 3.3

SAS Infrastructure for Risk Management 3.4

Example Code and Data

You can access the example code and data for this book by linking to its author page at https://support.sas.com/bequet. Larger versions of some flow diagrams are also available on this page.

We Want to Hear from You

SAS Press books are written *by* SAS Users *for* SAS Users. We welcome your participation in their development and your feedback on SAS Press books that you are using. Please visit sas.com/books to do the following:

- Sign up to review a book
- Recommend a topic
- Request information on how to become a SAS Press author
- Provide feedback on a book

Do you have questions about a SAS Press book that you are reading? Contact the author through saspress@sas.com or https://support.sas.com/author_feedback.

SAS has many resources to help you find answers and expand your knowledge. If you need additional help, see our list of resources: sas.com/books.

About The Author

Henry Bequet is Director of High Performance Computing and Machine Learning at SAS. In that capacity, he leads the development of a high-performance solution that can run SAS code on thousands of CPU and GPU cores for advanced models that use techniques like Black-Scholes, Binomial Evaluation, and Monte-Carlo simulations. Henry has more than 35 years of industry experience and 15 years of high-performance analytics practice. He has published two books and several papers on server development and machine learning.

Learn more about this author by visiting his author page at http://support.sas.com/bequet. There you can download free book excerpts, access example code and data, read the latest reviews, get updates, and more.

Acknowledgments

Writing a technical book is hard work for everyone involved. Completing the first draft is only the first step. To pass the finish line, you will need the support, the dedication, and above all the patience of all the people involved directly and indirectly in the project. I would like to take this opportunity to express my gratitude to all who helped me in this endeavor.

To my dedicated editor, Lauree Shepard, and forgiving reviewers, Juan Du, Huina Chen and Maged Tawfik, I want you to know how much I appreciated your input.

As I mentioned earlier, the genesis of this book was the design and implementation of the SAS Infrastructure for Risk Management platform, so it is only fair to acknowledge the contributions of my manager, Stacey Christian, who trusted me to lead the development of the SAS Infrastructure for Risk Management platform. Without the continuous support and dedication of my department, this book would not have been possible: many thanks to Kais Arfaoui, Partha Dutta, Ron Stogner, Karen Taylor, Eric Yang, and Ricky Zhang. I would also like to thank Oliver Schabenberger for agreeing to write the preface of this book and for the support that he gave to our team and our technology.

For this challenge and many others that I was foolish enough to undertake, success would have been elusive without the unconditional backing of my wife, Anne-Françoise: merci de tout cœur pour ton support.

Chapter 1: Introduction

Deep Learning

This is a book about deep learning, but it is not a book about artificial intelligence.

In the remainder of this introduction, we explain those two statements in detail with a simple goal in mind: to help you determine whether this book is for you.

Let's begin by briefly discussing deep learning (DL)—more specifically, its pros, cons, and applicability. Then we will discuss the main motivation of this book: execution speed of analytics. We will defer a discussion on the mechanics of DL to Chapter 2.

For our discussion, we view DL as a technology with a straightforward goal:

Build a system that can predict outputs based on a set of inputs by learning from data.

You will notice that there are absolutely no references to a human brain, cognitive science, or creating a model of human behavior in this book. DL can do all those things and can do them very well, but that is not the focus of this book. For this book, we simply concentrate on creating a model (or building a system) that can predict outputs with some level of accuracy, given some inputs.

Like many technologies (some might argue any technology), DL has its advantages and disadvantages. Let's start with the advantages to keep our motivation high in these early stages.

Here are three of the main advantages of DL:

- DL provides the best performance on many data-driven problems. In other words, DL provides the best accuracy and the fastest results. That is a bold claim that has been proven mathematically in some cases and empirically in many others. We investigate this bold claim in more detail in Chapter 3.
- DL provides great model and performance portability. A DL network developed for one problem can often be applied to many other problems without a significant loss of accuracy and performance. We see vivid examples of this portability in Chapters 3 and 7.

- DL provides a high level of automation of your model. Someone with good DL skills but little domain knowledge can easily create state-of-the-art models. Chapter 7 illustrates how powerful that characteristic is for modeling random walks.

These key advantages come at a cost:

- DL is computational and data intensive. Without a lot of both computational power and data, the accuracy of your DL models will suffer to the point of not being competitive.
- DL will not give out its secrets. This is true during training, where specifying the correct parameters is an art more than a science. This is also true during inference (a term that we define more clearly in Chapter 2). As you might already know and as we will show you in the remainder of this book, DL can give you great predictive accuracy for your models, but you cannot completely explain why it works so well.

Both of those disadvantages can be crippling, so let's discuss them further to help you determine their impact on your problems.

Is Deep Learning for You?

Computing resources during training was a crippling factor for neural networks during the last decade of the 20th century: the computing power wasn't available to train any but the simplest networks. Note that the term "deep learning" hadn't been coined yet; it most likely originates from the reviews and commentary of Hinton et al. (2006). The availability of computing resources is becoming less of a problem today thanks to the advent of many-core machines, graphics processing unit (GPU) accelerators, and even hardware specialized for DL.

Why is DL hungry for computing resources? Simply put, it's because DL is a subfield of computer science, and computer science thrives on computational resources. Without access to a lot of computational resources, you will not do well with DL. How much is a lot? Well, it depends, and we give some guidelines in quantifying computing resources in Chapters 4 and 8.

The fact that DL requires a lot of data for training is significant if you don't have the data. For example, if you're trying to predict shoppers' behaviors on an e-commerce website, you are likely to fail without accurate data. Manufacturing the data won't help in this case, since you are trying to learn from the data. Note that having an algorithm to manufacture data is a good sign that you understand the data. There are many other examples where a lot of data has made things possible and the absence of data is a crippling obstacle (Ng 2016).

The examples that we use in this book don't suffer from this drawback. When we don't have the data, we can manufacture it. For example, if we are trying to improve upon Monte Carlo simulations, as we do in Chapter 5, and we discover that we need a larger training set, there is nothing to worry about. We can simply run more Monte Carlo simulations to generate (manufacture) more data. In Chapter 6, we introduce one of the most powerful tools in the arsenal of the data scientist to produce a lot of training data: the general purpose graphics processing unit (GPGPU), or simply the graphics processing unit (GPU).

It's All about Performance

In the remainder of this introduction, we focus on speed, which is the main focus of this book. By now you must have decided that you can live with the drawbacks of DL that we just discussed. So you have enough data, have plenty of computing resources, and can live with the black box effect (the fact that DL doesn't give out its secrets) that often worries statisticians (Knight 2017).

If you're still on the fence, maybe the performance argument will convince you one way or another.

Flynn's Taxonomy

Most of the work presented in this book finds its roots in the Financial Risk Division at SAS. Financial institutions use a large number of computations to evaluate portfolios, price securities, and financial derivatives. Time is usually of the essence when it comes to financial transactions, so having access to the fastest possible technology to perform financial computations with enough accuracy is often paramount.

To organize our thinking around numerical application performance, let's rely on the following categories from Flynn's taxonomy (Flynn 1972):

- **Single instruction, single data (SISD)**
  A sequential computer that exploits no parallelism in either the instruction or data streams.
- **Multiple instruction streams, multiple data streams (MIMD)**
  Multiple autonomous processors simultaneously executing different instructions on different data.
- **Single instruction stream, multiple data streams (SIMD)**
  A computer that exploits multiple data streams against a single stream to perform operations that might be naturally parallelized.

Figure 1.1 shows Flynn's taxonomy on a timeline with the technologies associated with each classification (for example, GPUs are for SIMD). The dates and performance factors in Figure 1.1 are approximate; the main point is to give the reader an idea of the performance improvements that can be obtained by moving from one technology to another. As you will see as you read this book further, the numbers in Figure 1.1 are impressive, yet very conservative.

Figure 1.1: Performance of Analytics

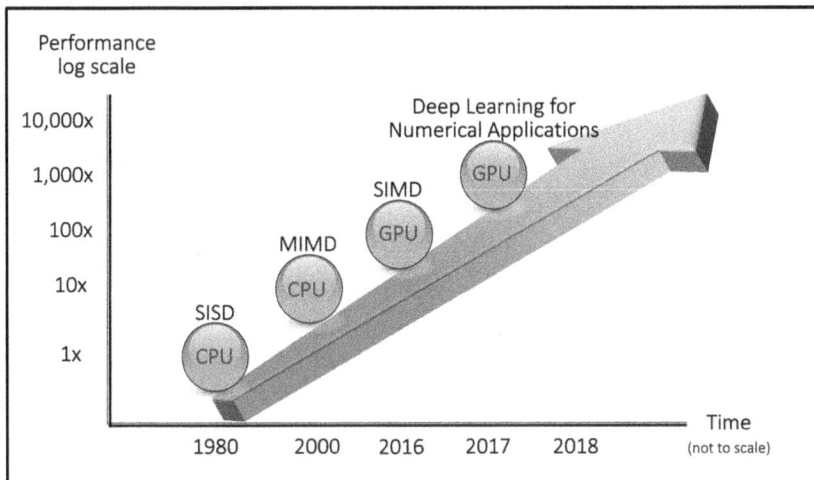

Life after Flynn

We start our exploration of the performance of numerical applications around 1980, when systems such as SAS started to be widely used. The SAS system (`sas.exe` that still exists today) is a SISD engine: SAS runs analytics one operation at a time on one data element at a time. Over the years, multi-threaded functionality has been added to SAS (for example, in PROC SORT), but at its heart SAS remains a SISD engine.

From the year 2000 to 2015 or so, analytics started to go MIMD with multiple cores and even multiple machines. Systems such as the SAS Threaded Kernel, the SAS Grid, Map-Reduce, and others gave folks access to much improved performance. We chose to give MIMD a 10x in our chart, but its performance was often much greater.

MIMD systems had and still have two main challenges:

- Make it as easy as possible to distribute the work across multiple cores and multiple machines.
- Keep the communication between the cores and the machines as light as possible.

As of this writing, finding good solutions to those two challenges still consumes a great deal of energy in the industry, and new products are still introduced, such as SAS Viya and SAS Infrastructure for Risk Management, to name only a couple. In terms of performance, the progress being made in the MIMD world is incremental at this point, so to go an order of magnitude faster, a paradigm shift is needed.

That paradigm shift comes in the form of the general purpose graphics processing unit (GPGPU), or simply graphics processing unit (GPU). GPUs are SIMD processors, so they need SIMD algorithms to process. To run quickly on GPUs, many algorithms have been redesigned to be implemented as SIMD algorithms (Satish et al. 2008). For example, at the time of this writing, most problems that occupy financial risk departments have a SIMD implementation. The most notable counter-examples are reports and spreadsheets. Potentially every single cell in a spreadsheet or a report implements a different formula (algorithm). This makes the whole report or spreadsheet ill-suited for SIMD implementations.

This last observation about reports and spreadsheets brings up an important point: as one moves up in our chart in Figure 1.1, not all problems can be fitted into the upper bubbles. Roughly speaking, any computable problem can be implemented with a SISD algorithm, a clear majority of the computable problems can be implemented with a MIMD algorithm, and a great number of problems can be implemented with a SIMD algorithm. One could visualize this applicability of algorithms to problems as an inverted cone. At the top of the cone (in the wide part), you find all applications that run on a computer, including yours. As you move down the cone, the number of applications shrinks, but at the same time the performance goes up. In other words, the closer to the bottom of the cone, the faster your application, but the less likely you are to find your application. As time goes by and new algorithms are developed, the narrow (bottom) tip of the cone becomes wider and wider.

But SIMD is not the final answer to fast performance for analytics; it is the beginning of the endeavor that we describe in this book.

We believe that the next paradigm shift with respect to the performance of numerical applications will come from deep learning. Once a DL network is trained to compute analytics, using that DL network becomes dramatically faster than more classic methodologies like Monte Carlo simulations. This latest paradigm shift is the main topic of this book.

Organization of This Book

This is a practical book: we want you to be able to reproduce the sample on your hardware with Base SAS and SAS Studio. You will not get the same results as what we publish in the book if you don't have the same hardware as what we used (who knows, yours might be faster!), but you will obtain similar results. To get the most out of the book, we advise you to follow the examples along with the book.

In Chapter 2, "Deep Learning," we provide a practical introduction to DL by describing the Deep Learning Toolkit (TKDL) that is available to SAS users. We start with a simple example of a cognitive application and then discuss how DL can go beyond cognitive applications.

Going beyond cognitive applications is precisely what we will do in Chapter 3, "Regressions." In that chapter, we show how the reader can use SAS in an application of the universal approximation theorem.

In Chapter 4, "Many-Task Computing," we take a slight digression from DL into supercomputing to introduce scalable deep learning techniques. In this chapter, we also discuss data object pooling, a technique that high-performance computing uses more and more to dramatically accelerate daily analytics computations. Chapter 4 provides one of the pillars of the foundation of the rest of book (the other pillar is DL).

In Chapter 5, we study Monte Carlo simulations. We begin with a simple deterministic example and then we progress to a stochastic problem.

In Chapter 6, "GPU," we leverage the awesome SIMD power of GPUs to manufacture extensive training data for a DL network.

In Chapter 7, "Monte Carlo Simulations with Deep Learning," we study how Monte Carlo simulations can be approximated using DL. The main takeaway from this chapter is that with a limited understanding of a domain and good DL skills, you can implement state-of-the-art analytics, both in terms of accuracy and in terms of performance.

In Chapter 8, "Deep Learning for Numerical Applications in the Enterprise," we describe how to gradually introduce deep learning for numerical applications into enterprise solutions. The main goal of this chapter is to convince you that the technologies described so far can be used to introduce an evolution to deep learning for numerical applications, not a revolution. We also discuss the best practices and pitfalls of scalability for deep learning.

Finally, in Chapter 9, "Conclusions," we summarize why deep learning for numerical applications is a powerful technique that allows SAS users to marry traditional analytics and deep learning to their existing analytics infrastructure. We also briefly discuss specialized hardware that will quickly become a viable solution because of the universality of DL.

But let's not get ahead of ourselves; we first need to look at the basics of DL and how to implement DL with SAS.

Chapter 2: Deep Learning

In this chapter, we introduce deep learning (DL). After looking at the history of DL, we then examine some concrete examples with SAS for a logistic regression (also known as classification). In the next chapter, we focus more on the type of regressions that are useful to accelerate numerical applications.

Deep Learning

In this section, we briefly discuss the history and the mechanics of DL. If you're already familiar with DL, feel free to skip this section and jump to the next section, "A Few Words about CAS."

In the following paragraphs, we put the emphasis on the technologies that are relevant to deep learning for numerical applications (DL4NA). It is the topic of this book after all. For a more complete and in-depth introduction to DL, please see Goodfellow et al. (2016).

Connectionism

Connectionism can be loosely defined as a technique that views a phenomenon as the result of the execution of processes of *interconnected networks of simple units*. A well-known example of such an interconnected network of simple units is the artificial neural networks (ANNs) that we use in DL.

The earliest reference to a network of connected units to reproduce some cognitive behavior dates back at least to the 19th century (James 1892). In that early case, the network was presented as an associative memory device (a device with content-addressable memory as opposed to a pointer-addressable memory that you find in most computers).

In the 1940s, Donald Hebb introduced the concept of *interconnected networks of simple (computational and memory) units* (Elman et al. 1996). During the same period, in 1943, Warren S. McCulloch and Walter Pitts published their landmark paper on the cognitive process (McCulloch and Pitts 1943). In their paper, McCulloch and Pitts gave a highly simplified model of the neurons in the mammal brain. At the time, the existence of neurons and some of their behaviors were understood. However, McCulloch and Pitts were trying to understand how assembling many neurons can lead to a complex cognitive process, namely intelligence. In 2018, it is not clear that we have a good solution to the problem that McCulloch and Pitts were trying to solve back in 1943, but we have to thank them for the concept of an idealized neuron that can be assembled into a large network of neurons to learn from data. That concept is at the core of DL.

The Perceptron

About a decade later, Frank Rosenblatt had the idea of building a machine to classify images (Rosenblatt 1957). The perceptron was born. More specifically, the *single layer* perceptron was born. As we will see shortly, the distinction between single and multiple is crucial.

A perceptron is what we call today a linear binary classifier. A perceptron implements the following function:

$$f(x) = \begin{cases} 1, & w \cdot x + b > 0 \\ 0, & otherwise \end{cases}$$

where x is a vector of the inputs, w is a vector of weights, b is a vector of the biases, $\cdot$ is the dot product $\sum_{i=1}^{n} w_i x_i$, and n is the number of inputs to the perceptron. The inputs, x, are usually called features.

Graphically, the situation is represented in Figure 2.1.

Figure 2.1: The Perceptron

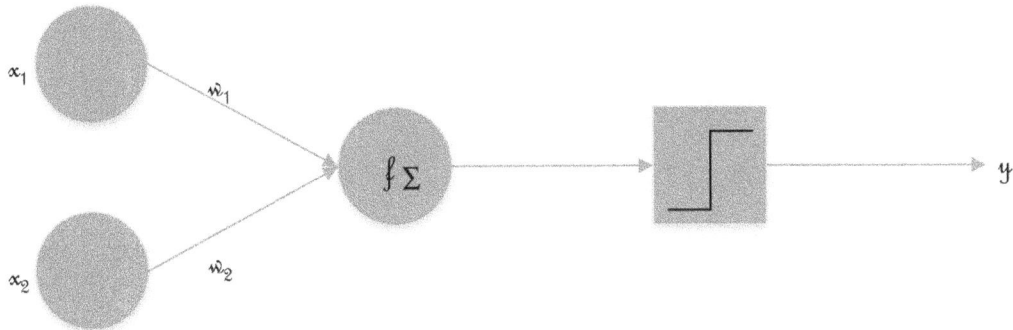

Notice the step activation function (in the box) that returns either 0 or 1. We will see more practical activation functions shortly.

Since the output of f is either 0 or 1, tweaking the values of w and b allows us to classify the inputs into two classes: the class that returns 1 versus the class that returns 0. Looking at the preceding formula, we can quickly infer that the value of the bias allows us to move the *decision boundary*: a low bias allows the perceptron to fire (return 1) for small values of $w \cdot x$, and conversely, a high bias makes it harder for the perceptron to fire (hence, the name "bias"). Another way to look at the bias is to state that the bias is the simplification assumption made by the perceptron to make it easier to reach a satisfactory approximation of the target.

This simple observation on the effect of the bias is important. It implies that the perceptron will never be able to correctly classify a training set that is not linearly separable. In a two-dimensional space, this is equivalent to stating that a perceptron can correctly classify the elements of a training set with a class for stars and a class for crosses, as you can see in Figure 2.2. In that figure, the decision boundary that separates the crosses from the stars is a line.

Figure 2.2: Linearly Separable Classes

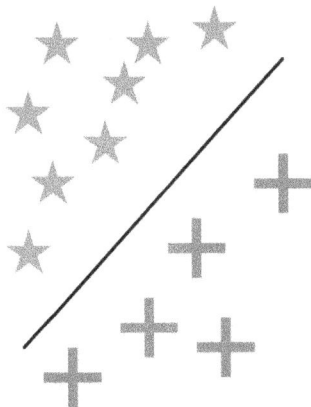

But a perceptron will never be able to classify the training set of Figure 2.3 with a satisfactory level of accuracy.

Figure 2.3: Non-linearly Separable Classes

One point should be noted: In that last training set, using a polar coordinate system would have resulted in a linearly separable training set that the perceptron would have classified correctly. In other words, in this case, the representation of the features can make or break a simple classifier like the perceptron. Choosing a good coordinate system is sometimes called feature representation. Selecting the correct features and then a good representation for those features is an essential step in most machine learning projects, but with the advent of DL, this step is rarely a make-or-break decision. Most of the time, a poor choice of features results in a network that is hard to train. Like many aspects of machine learning (ML), this important step of feature selection and representation is more art than science. We will see an example of feature representation that can make a significant difference when we look at the classification of irises later in this chapter.

So, how do you get good values for *w* and *b* after you have selected good features with a good representation? The answer is simple but profound:

> *You learn from the data with a learning algorithm that iterates until it reaches a satisfactory approximation.*

You don't program or set the values of *w* and *b* by hand or via code (at least not directly; there is always some code involved). In other words, *the programming is automatically done by the learning algorithm*, not manually by a C programmer or a SAS developer. The words "satisfactory approximation" are very important. A learning algorithm is not exact; in fact, in all the examples that we use in this book, the learning algorithm is stochastic. (A learning algorithm is sometimes called a training algorithm, and we will use those terms interchangeably.)

The first training algorithm for the perceptron was proposed by Frank Rosenblatt, and its implementation on the Mark I Perceptron machine was a success. In fact, it was perhaps too much of a success. In the euphoria that followed the Mark I, Rossenblatt made exaggerated claims that were echoed in the press to the point that it became common belief that the perceptron was going to be used to build a machine that would be self-aware (of its existence) (Olazaran 1996).

The First AI Winter

As you know, a self-aware perceptron failed to materialize. Furthermore, when Marvin Minsky and Seymour Papert (1969) showed in 1969 that a single perceptron was incapable of implementing a simple logical operation like XOR, the first artificial intelligence (AI) winter was about to begin, and connectionism was about to be relegated to the bin of failed technologies, at least for the duration of the first AI winter.

So, what is an AI winter? An AI winter is a period of reduced funding and interest in AI projects. Typically, an AI winter comes after a period of hype with unreasonable expectations around an AI technology such as the perceptron. Essentially, the hype either gets ahead of the theory or the technology or both.

The first AI winter started shortly after the disillusionment around the perceptron. By some accounts, it lasted from 1974 to 1980.

The Experts to the Rescue

So, what happened around 1980 to start an AI spring? Expert systems are what renewed the interest and the hype in the early 1980s. The idea was simple: program a machine by entering a set of rules (knowledge base) and let a program (called an inference engine) organize the rules to reach a conclusion. During the 1980s, several languages and inference engines were introduced. LISP and PROLOG are probably among the most famous. Alas, those systems didn't scale at all, neither in relation to problem complexity nor performance.

The failure to address complex problems was most likely due to the lack of machine learning in those systems. The process was more like "machine teaching" by teachers who couldn't formally codify why they would reach complex conclusions. In hindsight, this methodology was doomed from the start.

The Second AI Winter

On the raw performance side, the machines of the 1980s were just not capable of handling the combinatorial explosions of the inference engines. So, with the collapse of the market for the LISP machines in 1987, the second AI winter was ready to start.

The second AI winter lasted from 1987 to at least 1993.

One paper that was published before the second AI winter started, in 1986, deserves a special mention: "Learning representations by back-propagating errors" (Rumelhart et al. 1986). Geoff Hinton, who we will mention several times in the coming pages, was among the authors of that revolutionary paper. In that paper, the authors laid down the foundation of the DL algorithms that are widely used today: rely on an optimizer (gradient descent) to minimize the errors of the network by adjusting the weights and biases. In our programming, we use a variation of the gradient descent method, adaptive moment estimation, which is also known as the Adam optimizer (Kingma and Ba 2015).

The Deeps

By 1993, people were very cautious when it came to AI: twice burned, thrice shy! So the decade from 1993 to 2006 saw little irrational exuberance around AI. A few examples of carefully crafted projects were mostly fueled by Moore's Law. For example, in 1997 Big Blue became the first computer to defeat a grand master (Garry Kasparov) in chess. This victory was mostly due to ingenious programming and raw computing power: How quickly can you navigate a tree of all the possible moves? Consequently, there was one Big Blue and that was it. The technology couldn't be replicated for other AI problems such as computer vision. This kind of breakthrough was still more than a decade away. Here again, there was no machine learning. Instead, there were many smart programmers who were hand-coding the search and rank algorithm that would run on one of the fastest super-computers of the time.

Things started to get more interesting when Geoff Hinton proposed the deep belief network along with a fast training algorithm (Hinton et al. 2006). Using this strategy, people were able to train deeper networks of perceptrons: the multilayer perceptron (MLP) networks that we use throughout this book.

By increasing the number of perceptrons and by using more than a single layer, people were moving closer to DL, but were not quite there yet. Training was too slow and accuracy was too low to rival other technologies. A few key innovations were still required to reach the DL that we know today.

During the early 2010s, several of these key innovations appeared. The first notable one was the introduction of the rectified linear unit (ReLU) as the activation function (Nair and Hinton 2010). Prior to the ReLU, the sigmoid function was a popular choice. However, the derivative of the sigmoid is always less than 1.0, and as you increase the number of layers and neurons, the product of the derivatives quickly approaches 0, which gives you little to work with to adjust the weights and biases (in addition to the numerical instabilities associated with very small numbers). This problem is called the vanishing gradient. The ReLU function doesn't suffer from the vanishing gradient problem, so using the ReLU as an activation function allowed the training to converge faster.

As practitioners were able to better train large networks thanks to innovations like the ReLU, they started to encounter another problem: overfitting. Overfitting occurs when your network learns the training set a little too well and fails to generalize. In practical terms, the error is small during training, but shoots up during testing when you feed your network with a completely new set of data (for example, new images). To reduce the likelihood of overfitting, Geoff Hinton introduced the dropout method (Hinton et al. 2012).

These enhancements were important, but the real breakthrough and the birth of DL came in 2012, during an image classification competition. Computer science was about to give DL the extra push that it needed to become competitive.

The Large Scale Visual Recognition Challenge (LSVRC) is a yearly competition. During the LSVRC, the best minds in the computer vision field compete to build models that can recognize as accurately as possible the millions of labeled images stored in the Imagenet database (Deng et al. 2009). The error rates of the winners for the first three years of the LSVRC competition were as follows:

Table 2.1: Error Rate for Three Years of the LSVRC Competition

Year	Error Rate
2010	28%
2011	26%
2012	16%

The 2012 winning entry submitted by Alex Krizhevsky, Ilya Sutskever, and Geoff Hinton was an unprecedented breakthrough in computer vision, especially if you consider that the authors were not computer vision experts, but ML experts. The 2012 winning entry was the only entry that relied on DL.

What happened? Why was this dramatic improvement possible? In one word: GPUs.

The model was trained for about a week on two NVIDIA GTX 580 GPUs. As we will see in Chapter 6, graphics processing units or GPUs provide data scientists with thousands of processors to run their programs. Prior to the availability of GPUs, only a few processors were available (for example, 8 or 16). With orders of magnitude of increases in the availability of raw computing power, the training of deep neural networks was becoming practical.

In addition to the awesome power of the GPUs, the deep neural network used to win the LSVRC had several mainstays of the networks that we use today, namely ReLU as an activation function and dropout as a regularization method.

Now that we have DL, is history about to repeat itself? Are we at the dawn of the third AI winter?

The Third AI Winter

A third AI winter is always possible, but as of this writing, it is unlikely for several reasons.

The progress that has been made since 2012 has been sustained and significant. In addition, from AI that wins all games of Go to image recognition to automatic translation to self-driving cars, the reach of AI has been broad and deep. This was not the case in the past. The first AI winter came after the disappointments of the one and only perceptron and the lack of generalization of the one and only computer vision application. The second AI winter came after the collapse of the LISP market. Of course, there were other symbolic systems, but certainly not a plethora comparable to the explosion of applications around ML and AI that we see today (this book describes one of them).

The data aspects of this equation should not be underestimated. ML learns from data: the more, the better. If the training data were not available, that would definitely be a concern. But big data is not slowing down, it is accelerating. "IDC forecasts that by 2025 the global data sphere will grow to 163 zettabytes (that is a trillion gigabytes or 10^{21} Bytes). That's ten times the 16.1 ZB of data generated in 2016" (Reinsel et al. 2017). All this data will unlock an avalanche of ML opportunities. This is abundance of data, not scarcity of data. This is abundance of fuel for ML.

In addition to the advances that we've seen on the data and the software side, the advances on the hardware side have been nothing short of amazing. The GTX 580 was in the GFLOPS range (10^9 floating point operations per second). Today's GPUs (only 5 years later) routinely operate in the TFLOPS range (10^{12}), and clusters of GPUs are approaching the EFLOPS range (10^{18}). And this is by no means the limit, as we will discuss in Chapter 9. Hardware vendors have roadmaps that will take the performance of ML and AI orders of magnitude faster than what you see today.

In summary, the lack of applications and adequate hardware that fueled the previous winters don't seem to apply to the current situation. As of this writing, the best is yet to come in DL!

Some Supervision Required

Our focus in this book is to leverage the power of DL to gain significant performance improvements in numerical applications. This is to say that, typically, we already know what we need to compute and, at least during the development phase, we know the actual results that we seek. This is the quintessential definition of *supervised learning*: before training the ANN, the observations are already labeled with a class or a target result.

In this book, we exclusively focus on supervised learning.

There are several other types of learning that we ignore in this book. We briefly mention a few for awareness:

- Unsupervised learning

  In unsupervised learning, the training data is not labeled a priori and it is up to the training algorithm to figure out what needs to be learned. The textbook example of such a learning is clustering.
- Reinforcement learning

  In this case, the desired outputs are never taken into consideration, but the learning algorithm tries to maximise a reward.
- Adverserial learning

  This type of learning is a generalization of reinforcement learning where two or more classifiers have different loss functions.

At this point, we have enough information to create ANNs, train ANNs, and run predictions with ANNs. However, we need to say a few words about the ANN engine that we will use, SAS Cloud Analytic Services (CAS).

A Few Words about CAS

In this section, we briefly present CAS, which we use for most of our DL work throughout the book. This introduction is by no means comprehensive. We only focus on the small subset of CAS features that we need in this book.

If you're already familiar with CAS, please feel free to skip this section and jump to the next section, "All about the Data."

An in-depth presentation of CAS can be found in the SAS documentation (A Guide to the SAS 9.4 and SAS Viya Programming Documentation located here: https://support.sas.com/documentation/onlinedoc/viya).

Deployment Models

CAS provides the high-performance processing power of SAS Viya. SAS Viya and CAS supplement the features that are available in SAS 9.4. As you will see throughout this book, there is a smooth integration between SAS 9.4 and SAS Viya that we repeatedly leverage.

As its name suggests, CAS can be deployed in the cloud, but it can also be deployed on premises. When deployed at your site, the CAS server can run on a single machine or as a distributed server on multiple machines:

- The single-server deployment offers a symmetric multi-processing (SMP) environment.
- The distributed server comprises one controller and one or more workers that communicate using a protocol called Message Passing Interface (MPI) to provide a massively parallel processing (MPP) environment.

Figure 2.4 shows a diagram of a distributed server with the controller and the workers.

Figure 2.4: CAS Processes (Source: SAS Institute)

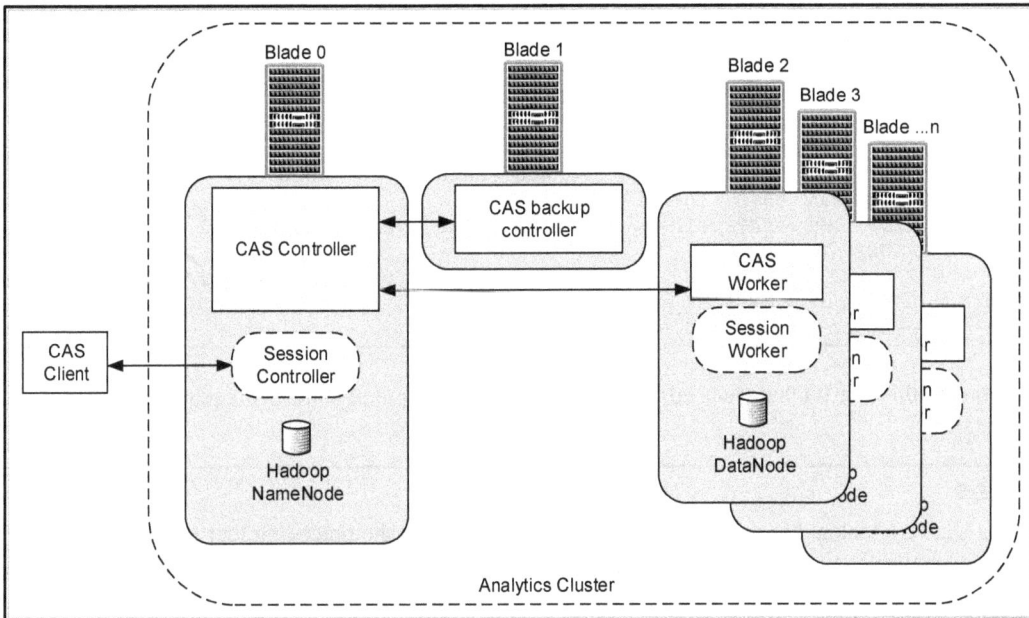

In both deployment models, the threaded processing of the CAS server is shared by multiple CPUs or GPUs. In the MPP case, the sharing of processing happens on multiple machines. GPU processing is not available in all cases, but we will make a point of using GPU processing when it is available.

In SMP mode, SAS 9.4 and SAS Viya can be deployed on the same machine. That is the deployment that we use in this book.

As a programming interface to CAS, we mostly rely on SAS Studio.

CAS Sessions

To run any anything in CAS, you must log in to CAS. The log-in process establishes a CAS session that you must use to submit any request to CAS. In practice, starting a session comes naturally because CAS maintains the concept of a current session: once you've logged in successfully, the CAS session that you obtain is the current session, and any request to CAS works with that (current) session unless you specify otherwise.

Let's look at a quick concrete example. We first log in to SAS Studio as the user dl4na and then run the following code:

```
options cashost='localhost' casport=5570;
cas mysession;
```

The `options` statement specifies where the CAS server can be found (server and port). As we stated earlier, we deployed CAS on the same machine as SAS 9.4 where SAS Studio is installed, so the machine is `localhost`. The port number was specified during installation. In this case, we used the default. It is perfectly fine to use another machine to deploy CAS. It is also perfectly fine to use a grid to deploy CAS and run SAS Studio on an SMP machine.

We then start a CAS session. That might seem odd: What credentials do we use to start the new CAS session? Thanks to the seamless integration between SAS 9.4 and CAS, we use the same credentials that we used to log in to SAS Studio. This becomes obvious when you examine the SAS log:

```
  1              OPTIONS NONOTES NOSTIMER NOSOURCE NOSYNTAXCHECK;
 80             options cashost='localhost' casport=5570;
 81             cas mysession;
NOTE: The session MYSESSION connected successfully to Cloud Analytic Services
localhost using port 5570. The UUID is
        7cbd4c0a-5a98-4f47-b61b-e44de180c634. The user is dl4na and the active
caslib is CASUSER(dl4na).
NOTE: The SAS option SESSREF was updated with the value MYSESSION.
NOTE: The SAS macro _SESSREF_ was updated with the value MYSESSION.
NOTE: The session is using 0 workers.
```

In addition to the user ID that we used to connect to CAS, a couple of concepts require further explanations.

Caslibs

First, let's have a look at the concept of a caslib, as mentioned in the preceding log:

```
the active caslib is CASUSER(dl4na).
```

The main purpose of a caslib is to provide an in-memory space for CAS to perform operations on tables.

A caslib is automatically created when you log in. That automatically created caslib is called the personal caslib and is analogous to the SASUSER libref in SAS 9.4, although there is no concept of in-memory for a libref. Since we logged-in as dl4na, our personal caslib is called CASUSER(dl4na). We can ask CAS to show us more information on the personal caslib:

```
options cashost='localhost' casport=5570;
cas mysession;

caslib casuser list;
```

The caslib statement code produces the following log:

```
 83             caslib casuser list;
NOTE: Session = MYSESSION Name = CASUSER(dl4na)
        Type = PATH
        Description = Personal File System Caslib
        Path = /home/dl4na/casuser/
        Definition =
        Subdirs = Yes
        Local = No
        Active = Yes
        Personal = Yes
```

As you undoubtedly expected, the personal caslib is flagged with Personal = Yes. You'll also notice a path (/home/dl4na/casuser) that is used in case you want to save the content of your tables.

Using the CAS LIBNAME engine, you can easily transfer data from SAS 9.4 to CAS. For example, the following statements would copy the `SASHELP.iris` data set from the 9.4 `SASHELP` libref to the active caslib, in our case `CASUSER(dl4na)`:

```
libname INPUT cas;

/* Load the training data set into CAS */
data INPUT.iris;
   set SASHELP.iris;
run;
```

The SAS 9.4 and SAS Viya programming documentation contains much more information about caslibs, but what we have described here is enough for this book.

Workers

The preceding log references a worker:

```
NOTE: The session is using 0 workers.
```

So, what is a worker in the context of CAS? It is a machine, a physical server. Figure 2.4 shows a representation of what happens on a grid deployment. In that diagram, we also reference a controller, which can be described as the conductor of the work that needs to happen on all the workers.

In an SMP deployment (like the single machine we use in this book), the number of workers is always 0, since the controller is doubling up as a worker. Everything happens in one process in SMP mode. That doesn't mean that everything happens in one thread of execution. Thanks to many CPU cores and GPUs, we can run our DL algorithm in thousands of parallel threads of execution on a single machine.

Action Sets and Actions

An action is the way you get things done in CAS. For convenience, actions that perform similar operations are typically grouped into action sets. For instance, there is an action set for simple statistics such as a distinct count or the computation of the Pearson product-moment correlation.

How do you trigger the execution of the action in CAS? Using the CAS procedure (PROC CAS), the user (you) sends a request to the server in the form of an action with its parameters (for example, summarize a table or train an ANN). Then that request is parsed by the controller and the work gets distributed over all the workers or the SMP machine if only one server is involved. After the work is completed, the results come back to you, the user.

Let's see a simple example where we count the number of rows in the IRIS table that we loaded into CAS a few paragraphs ago:

```
proc cas;
  simple.distinct /
      table={name="IRIS"};
quit;
```

The log would look something like this, simply indicating that CAS executed the requested action (notice the reference to the active session):

```
94          proc cas;
95             simple.distinct /
96                table={name="IRIS"};
97          quit;

NOTE: Active Session now MYSESSION.

NOTE: PROCEDURE CAS used (Total process time):
      real time            0.06 seconds
      cpu time             0.06 seconds
```

The result is shown in Figure 2.5.

Figure 2.5: Results of a CAS Action

Distinct Counts for IRIS			
Column	Number of Distinct Values	Number of Missing Values	Truncated
Species	3	0	No
SepalLength	35	0	No
SepalWidth	23	0	No
PetalLength	43	0	No
PetalWidth	22	0	No

The first thing to notice is that we submit requests to the CAS server via PROC CAS, which uses the current CAS session and the current caslib: the IRIS table was loaded from the INPUT caslib (because of the libname INPUT cas statement).

We then invoked the simple.distinct action by sending it to the controller; the controller parsed the request, triggered the processing, and sent the result back to us for display in the log.

In the preceding example, the action set is `simple` and the action is `distinct`. As we mentioned earlier, the `simple` action set contains several actions aimed at computing simple statistics (only a few of the actions are shown):

Action Name	Description	Label
Correlation	Computes Pearson product-moment correlations.	Pearson product-moment correlation
crossTab	Performs one-way or two-way tabulations.	Tabulation
distinct	**Computes the distinct number of values of the variables in the variable list.**	**Distinct count**
freq	Generates a frequency distribution for one or more variables.	Frequency
groupBy	Builds BY groups in terms of the variable value combinations given the variables in the variable list.	Group-by

Cleanup

At the end of a SAS program that accesses CAS, it is a good idea to terminate the CAS session so that it no longer consumes resources in the controller and the workers.

Here is the complete listing of the `simple_cas.sas` program:

Program 2.1: The simple_cas.sas Program

```
options cashost='localhost' casport=5570;
cas mysession;

caslib casuser list;

libname INPUT cas;

caslib _all_ list;

/* Load the training data set into CAS */
data INPUT.iris;
   set SASHELP.iris;
run;

/* Perform some simple statistics */
proc cas;
  simple.distinct /
      table={name="IRIS"};
quit;

cas mysession terminate;
```

We now have enough information to use CAS in conjunction with SAS 9.4. If you would like a more in-depth description of SAS Viya and CAS, please refer to the SAS documentation (A Guide to the SAS 9.4 and SAS Viya Programming Documentation).

All about the Data

In this section, we touch on the importance of the data, or more specifically, the importance of looking at your data before embarking on a DL project.

For the remainder of this chapter, we use data sets that are available with any SAS installation. Those data sets are small, but the ones that we use are big enough to learn simple relationships and make predictions. When you're comfortable with the material presented in this chapter, feel free to experiment with larger data sets like the ones available in the UCI Machine Learning Repository: http://archive.ics.uci.edu/ml/index.php (you will find some of the SAS sample data sets in that repository).

The Men Body Mass Index Data Set

The SASHELP.bmimen data set contains the variables BMI and Age. The Body Mass Index (BMI) is defined by the ratio of the weight in kilograms to the square of the height in meters. Since BMI is widely used to classify individuals as underweight or overweight, we would like to start a DL project to examine how the age can be a predictor for these two classes in the population that we are investigating.

Given the fact that DL learns from data, we want to have a cursory look at the data to perform some sanity checking. So let's load the data into SAS and plot some simple graphs:

Program 2.2: Loading Data and Plotting Graphs

```
options cashost='fsnlax05' casport=5570;
cas mysession;

libname INPUT cas sessref = mysession;

/* Load the training data set into CAS */
data INPUT.bmimen;
   set SASHELP.bmimen;
   where age < 10;
run;

/* Have a look at the data we loaded into CAS */
proc casutil; list;
quit;

ods graphics / reset width=6.4in height=4.8in imagemap;

proc sgplot data=INPUT.bmimen;
   title 'Body Mass Index by Age for Men Data';
   histogram BMI /;
   yaxis grid;
run;

proc sgplot data=INPUT.bmimen;
   title 'Body Mass Index by Age for Men Data';
   scatter x=age y=bmi / ;
run;

ods graphics / reset;

cas mysession terminate;
```

This type of code must now seem routine: we start a session with CAS and then we load the data from SASHELP to CAS. You probably noticed that we didn't load all the data into CAS:

```
where age < 10;
```

Since we are going to plot the data using the table in CAS, it is a good idea to limit the amount of the data that we are working with.

To double-check that the data was correctly loaded into SAS, we use the CASUTIL procedure (PROC CASUTIL).

The result of PROC CASUTIL is shown in Figure 2.6. We see the personal caslib that we used (`CASUSER(dl4na)`) along with some information about the BMIMEN table. So far, so good—no surprises.

Figure 2.6: Result of PROC CASUTIL for the SASHELP.bmimen Data Set

Caslib Information	
Library	CASUSER(dl4na)
Source Type	PATH
Description	Personal File System Caslib
Path	/home/dl4na/casuser/
Session local	No
Active	Yes
Personal	Yes
Hidden	No
Transient	Yes

Table Information for Caslib CASUSER(dl4na)								
Table Name	Number of Rows	Number of Columns	Indexed Columns	NLS encoding	Created	Last Modified	Promoted Table	Repeated Table
BMIMEN	1515	2	0	utf-8	16Mar2018: 13:17:38	16Mar2018: 13:17:38	No	No

The results of the PROC SGPLOT statements are shown in Figure 2.7.

Figure 2.7: Results of PROC SGPLOT for the SASHELP.bmimen Data Set

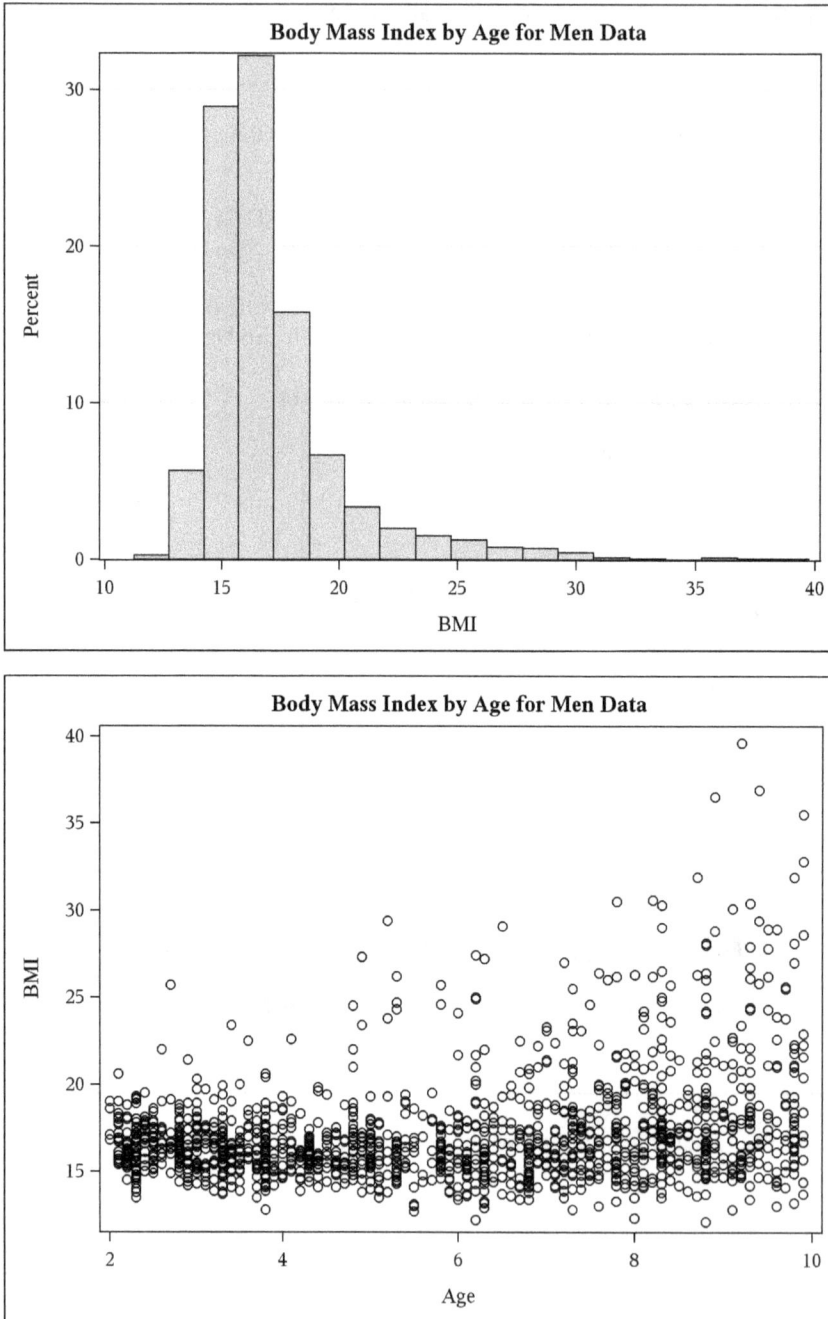

Body Mass Index by Age for Men Data

Body Mass Index by Age for Men Data

The histogram is no real surprise. We see a right-skewed distribution with a mean closer to 15 than to 20. No obvious signs indicate that something is amiss.

The scatter plot is another matter. From the plot, there is no way to accurately predict the BMI based on the age, at least in the population that we are studying (you will see a similar picture after 10 years of age). In other words, the age cannot be an accurate predictor of the BMI in the population that we are investigating.

This little example illustrates how important it is to know your data. It also illustrates that one of the best ways to look at your data is to plot it. We will see that during training, plotting the characteristics of the training, such as the loss function, is also very informative.

In case you're wondering, here's what an ANN would do with such a data set: In the best case scenario, it would estimate the mean of the training data with a mean square error (MSE) around 10. The MSE is a common measure of the error during the training that finds its origins in regressions. We will revisit this topic in Chapter 3.

In conclusion, the SASHELP.bmimen data set is not a good candidate to illustrate our discussion on deep learning, so let's look at another one and see whether it meets our needs better.

The IRIS Data Set

The IRIS data set (Fisher 1936) is an example of a multivariate data set widely used for examples of discriminant analysis and cluster analysis.

There are three species of irises in the 150 observations of the data set:

- *Iris setosa*
- *Iris versicolor*
- *Iris virginica*

The predictors for this classification are the millimeter measurements of the sepal length, sepal width, petal length, and petal width. There are 50 iris specimens from each of three species. Here is a sample of the measurements:

Table 2.2: Measurements in the IRIS Data Set

Species	Sepal Length	Sepal Width	Petal Length	Petal Width
Setosa	50	33	14	2
Setosa	46	34	14	3
Setosa	46	36	10	2
Setosa	51	33	17	5
Setosa	55	35	13	2
Versicolor	55	26	44	12
Versicolor	50	23	33	10
Versicolor	67	31	44	14
Versicolor	56	30	45	15
Versicolor	58	27	41	10
Virginica	60	22	50	15
Virginica	69	32	57	23
Virginica	74	28	61	19
Virginica	56	28	49	20
Virginica	73	29	63	18

Here again we would want to have at least a cursory look at the data, first by loading it to CAS and examining the output of PROC CASUTIL to make sure that there are no obvious surprises. The output of PROC CASUTIL is shown in Figure 2.8.

Figure 2.8: Output from PROC CASUTIL for the IRIS Data Set

Caslib Information	
Library	CASUSER(dl4na)
Source Type	PATH
Description	Personal File System Caslib
Path	/home/dl4na/casuser/
Session local	No
Active	Yes
Personal	Yes
Hidden	No
Transient	Yes

Table Information for Caslib CASUSER(dl4na)									
Table Name	Number of Rows	Number of Columns	Indexed Columns	NLS encoding	Created	Last Modified	Promoted Table	Repeated Table	View
IRIS	150	5	0	utf-8	16Mar2018:14:52:47	16Mar2018:14:52:47	No	No	No

As expected, we see 5 columns (the species + 4 measurements) and 150 rows (50 * 3 species). So far so good. So let's plot the three classes with the following code:

Program 2.3: Plotting Classes in the IRIS Data Set

```
ods graphics / reset width=6.4in height=4.8in imagemap attrpriority=none;

proc sgplot data=INPUT.iris;
   title 'Fisher (1936) Iris Data';
   styleattrs
     datasymbols=(circlefilled starfilled squarefilled);
   scatter x=petallength y=petalwidth / group=species ;
run;

ods graphics / reset;
```

The output of PROC SGPLOT is in Figure 2.9. You probably noticed that we switched the symbols (in addition to the colors) so that the scatter plot would be meaningful in black and white as well as in color.

We can see no obvious outliers and we also note that the measurements are within expectations, both for the sepals and for the petals.

Figure 2.9: Results of PROC SGPLOT for the IRIS Data Set

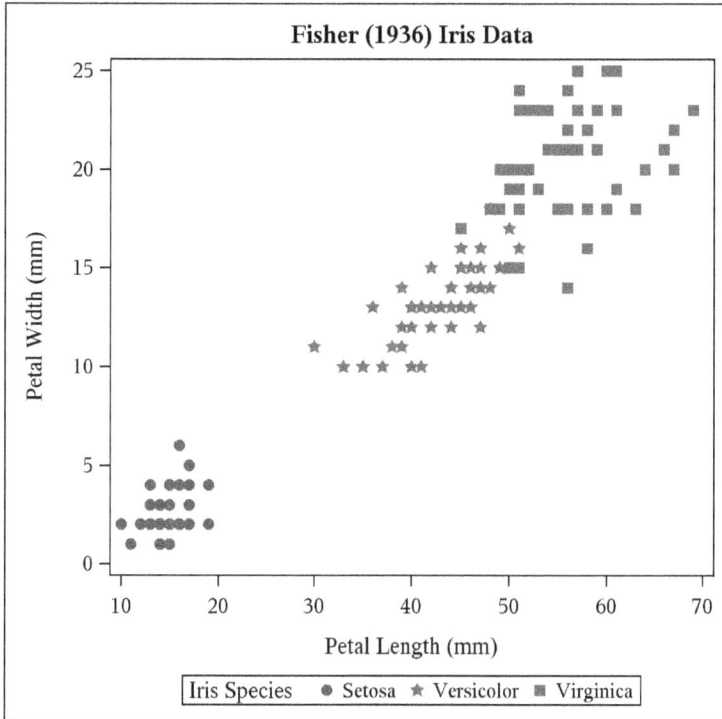

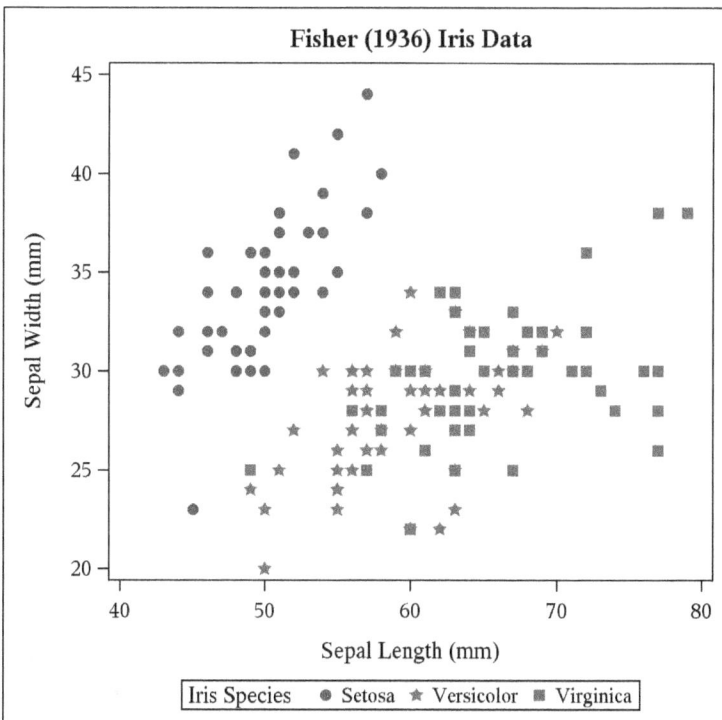

One thing should be obvious: this data set is by no means linearly separable, so a perceptron wouldn't do much with it. DL, however, would be able to easily learn from the data set and make very good predictions.

Before finishing up our cursory look at the data, we should point out that the sepal and petal lengths and widths are a choice of features. We could choose something else for features. For example, we could choose the area of the sepals and petals and get a much easier data set to classify, as shown in Figure 2.10.

Figure 2.10: Using Petal and Sepal Areas as Predictors (Image Credit: Rick Wicklin, 2012)

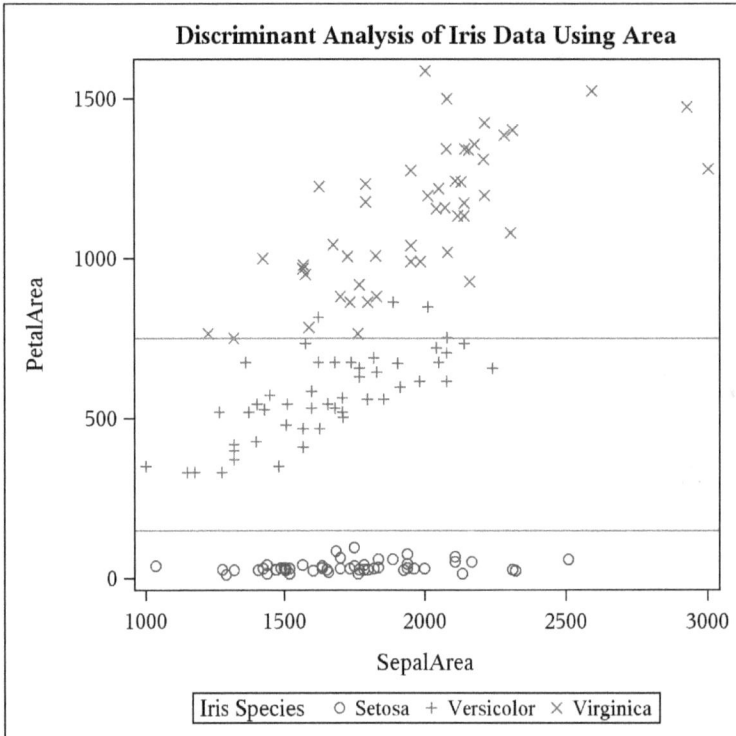

Now that we have a data set to demonstrate classification, let's build a model for it.

Logistic Regression

We will now build a deep learning model for the classification of the IRIS data set.

Preamble

As we discussed earlier, we will rely on the DL capabilities of CAS for our DL model. The use of CAS implies that we first need to create a CAS session as we did earlier.

Once we have a CAS session, we need to realize two things.

The first thing to realize is that the IRIS data set is sorted by species. That is not very good for training. To maximize our chances of converging quickly for all classes without overfitting, we need to (pseudo) randomly shuffle our input data set.

The second thing to realize is that we need three data sets, not one or two:

- One is for training, but we already knew that.
- One is for validation; that one is new.
- One is for testing; we already knew that as well.

The reason we need a validation data set is to avoid or at least detect overfitting for the training set. For a thorough discussion on the motivation around a validation set, see Hastie et al. (2009, 241–249), but the intuition is as follows.

During training, you keep a set of observations apart from the training set, and you call it the validation set. As we've mentioned before, the error on the training set is computed (on the training set) by the back-propagation algorithm to adjust the weights and biases. The observations in the validation set are not used to adjust the weights and biases; the observations of the validation set are only used as a predictor of the error on a test set. In the easiest possible use of the validation set, the training stops when the error on the validation set starts to increase (this is called early stopping).

Typical ratios between the training, validation, and test sets are 70/10/20 or 80/10/10. (The exact ratio doesn't matter that much; having the 3 data sets to do the job is the main point). So, in our case, we will create 3 (shuffled) data sets and load them into CAS with the following code:

Program 2.4: Shuffling and Loading Data Sets

```
options cashost='fsnlax05' casport=5570;
cas mysession;

libname INPUT cas sessref = mysession;

/* Shuffle the data */
data iris;
  set SASHELP.iris;
  sortval = rand('UNIFORM');
run;

proc sort data=iris out=iris(drop=sortval);
  by sortval;
run;

/* Load the training data set into CAS */
data INPUT.iris_train;
   set iris(obs=120);
run;

/* Load the validation data set into CAS */
data INPUT.iris_validation;
   set iris(firstobs=121 obs=130);
run;

/* Load the testing data set into CAS */
data INPUT.iris_test;
   set iris(firstobs=131);
run;
```

We have encountered all this code before with different data, so there is nothing surprising here. Things get more interesting when we start defining the DL model for training.

Create the ANN

Our goal is to create and then train a feedforward network like the one in Figure 2.11. We call it a deep neural network or DNN, since it is an ANN with more than one hidden layer.

Figure 2.11: A DNN for IRIS Classification

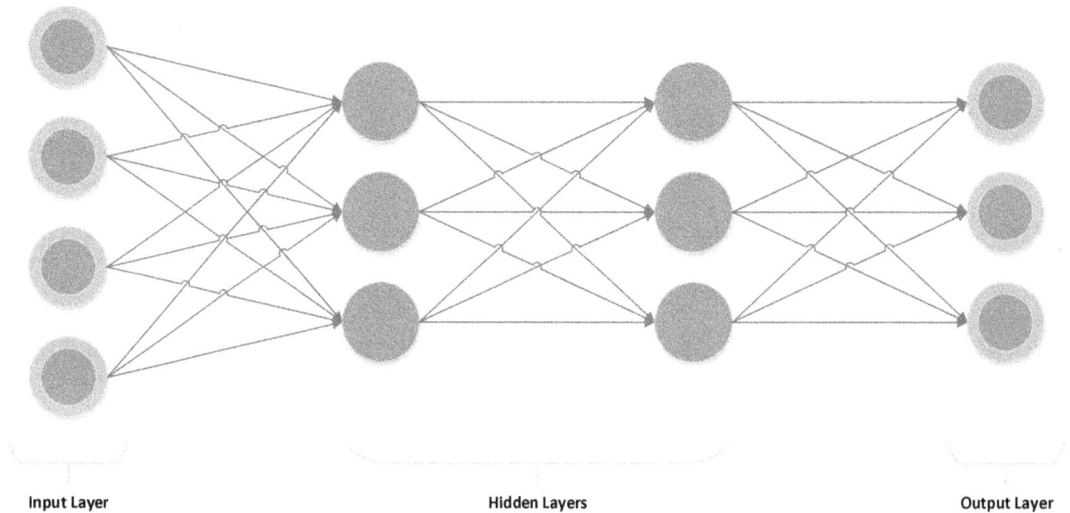

| Input Layer | Hidden Layers | Output Layer |

Our DNN has four layers: two visible layers and two hidden layers. The number of neurons of the input layer is driven by the number of features (the sepal and petal lengths and widths). The number of neurons of the output layer is driven by the number of classes that we want to recognize, in this case, three. The number of neurons in the hidden layers as well as the number of hidden layers is up to us: those two parameters are model hyper-parameters (more on this shortly).

Now that we have a good idea of what we're trying to build, let's get started.

The first thing to do to create the model is to invoke PROC CAS, then define some variables, and finally build the model using the `deepLearn.buildModel` action (`deepLearn` is the action set and `buildModel` is the action):

Program 2.5: Building the Model

```
proc cas; /* proc cas supports run-group processing */

/* First, we build the model */
model_name                 = "iris_logistic_regression";
training_table             = "iris_train";
validation_table           = "iris_validation";
testing_table              = "iris_test";
scoring_table              = "iris_score";
number_of_hidden_layers    = 2;
number_of_neurons_per_layer = 124;

deepLearn.buildModel/
  model = {name = model_name, replace = 1}
  type  = "DNN";
```

As you can see, PROC CAS supports the definition of variables, such as `model_name` and `number_of_hidden_layers`. The type of the variables is inferred: `model_name` is a string and `number_of_hidden_layers` is a number.

The names of the model and the tables are self-explanatory if you consider that classification and logistic regression are the same thing. However, two variables require a little explanation:

- `number_of_hidden_layers`
  We start with 2 hidden layers in our network.
- `number_of_neurons_per_layer`
  We start with 124 neurons in each (hidden) layer.

Those two parameters are called hyper-parameters because they are set before optimizing the model parameters (namely the weights and biases). We will see other examples of hyper-parameters later in this chapter. Since we give those model hyper-parameters a value, 2 and 124 respectively, you might be wondering how we picked those values. There is no easy answer: selecting hyper-parameters is more an art than a science. A good strategy is to start small (but not too small) and see how the training goes.

The arguments to `deepLearn.buildModel` simply define the name of the model and whether we should override an existing definition (the model definition goes into a CAS table by the same name). There is also a model type that we set to DNN (deep neural network). All the models in this book are DNN, which is a fully connected feedforward network (the ones that we've been describing so far). There are two other possible choices for the type of neural network at the time of this writing:

- convolutional neural network (CNN)
  CNNs are typically used for cognitive analytics like image recognition (the winning entry of the 2012 LSVRC competition was a CNN) and speech recognition. They find their origin in the work that David Hubel and Torsten Wiesel (1959) did on the visual cortex.
- recursive neural network (RNN)
  RNNs were developed in the 1980s by John Hopfield (1982). They are characterized by directed

cycles in the connections between the neurons and are well-suitable to model time series because of the temporal memory induced by the cycles.

Now that we have a few model hyper-parameters and an empty model, we need to start creating layers. The first layer is visible; it is the input layer that we add with the `deepLearn.addLayer` action:

Program 2.6: Creating the First Layer

```
deepLearn.addLayer/
  layer = {type = "INPUT"}
  name  = "input"
  model = {name=model_name};
```

This is straightforward. We simply add a layer of type `INPUT` to the DNN model that we just created, and we give that layer the name `input`. You might be surprised by what you don't see: the actual inputs. Don't worry, we will specify those once we are ready to train our network.

At this point, we have an input layer to feed our features to our network, but we still need to build the network, starting with the hidden layers:

Program 2.7: Adding Hidden Layers

```
previous_layer_name = "input";
do i = 1 to number_of_hidden_layers;
  new_layer_name = "hidden_" || i;

  deepLearn.addLayer/
    layer     = {type = "FC", act = "RELU",
                 n = number_of_neurons_per_layer}
    name      = new_layer_name
    srcLayers = {previous_layer_name}
    model     = {name = model_name};

    previous_layer_name = new_layer_name;
end;
```

A simple loop over the action provides us with what we need. You'll notice that we maintain a variable with the name of the previous layer to connect each layer to its predecessor. The following statement gives the values `hidden_1` and `hidden_2` to `new_layer_name`:

```
  new_layer_name = "hidden_" || i;
```

The `||` concatenates the string with the string representation of the loop variable `i`.

The type of layer is `FC`, which stands for a fully connected layer, since this is what we're building, as shown in Figure 2.11. The `srcLayers` parameter enables us to specify the name of the previous layer. In our case, there is only one predecessor because the network is fully connected. The `act` parameter is to specify the activation function. In our case, we want to use the ReLU because it worked so well for the LSVRC contest.

We are close to being done with building the network. We have an input layer and we have multiple hidden layers. An output layer that gives us the classification that we want is all that we're missing to complete our topology:

Program 2.8: Adding the Output Layer

```
deepLearn.addLayer/
   layer     = {type = "OUTPUT", act = "SOFTMAX", error = "ENTROPY"}
   name      = "output"
   srcLayers = {previous_layer_name}
   model     = {name = model_name};
run;
```

As you might have guessed, we still rely on the `deepLearn.addLayer` action and specify the type of the layer as being `output` (the last layer in our DNN).

The activation function is something new: `SOFTMAX`. This function enables us to satisfy two desirable conditions when running a DNN for classification:

1. The output of each neuron is going to be in [0, 1]
2. The sum of all the outputs is going to be 1.0

In other words, the output of the first neuron in the output layer can be interpreted as the probability that the inputs are in the first class. The second neuron can be interpreted as the probability that the inputs are in the second class, and so on.

In case you're curious, the softmax function is defined as follows:

$$y_c = \frac{e^{z_c}}{\sum_{d=1}^{C} e^{z_d}} \qquad \text{for } c \text{ in } 1, ..., C$$

where z is the vector of the outputs of the previous layer and C the number of classes.

Finally, we should say a word about the error or loss function: `ENTROPY`. When the output of a neural layer is interpreted as a probability, it is common and more correct to use the cross-entropy function instead of a squared Euclidian distance, for example. The cross-entropy is defined as follows:

$$L = -\sum_c y_c \log(z_c)$$

At this point, the specification of our DNN is complete, so we invoke `run` to look at the result, which you can see in Figure 2.12. Based on our previous discussions, none of this should come as a surprise. CAS gives an output for each action that we invoked and at each step we see an increase in the number of rows of the table that stores the model. Overall, this is a simple model that we can train quickly. And that's exactly what we're going to do in the next section.

Figure 2.12: DNN Model for Logistic Regression

Results from deepLearn.buildModel

Output CAS Tables			
CAS Library	Name	Number of Rows	Number of Columns
CASUSER(dl4na)	iris_logistic_regression	1	5

Results from deepLearn.addLayer

Output CAS Tables			
CAS Library	Name	Number of Rows	Number of Columns
CASUSER(dl4na)	iris_logistic_regression	10	5

Results from deepLearn.addLayer

Output CAS Tables			
CAS Library	Name	Number of Rows	Number of Columns
CASUSER(dl4na)	iris_logistic_regression	21	5

Results from deepLearn.addLayer

Output CAS Tables			
CAS Library	Name	Number of Rows	Number of Columns
CASUSER(dl4na)	iris_logistic_regression	32	5

Results from deepLearn.addLayer

Output CAS Tables			
CAS Library	Name	Number of Rows	Number of Columns
CASUSER(dl4na)	iris_logistic_regression	44	5

Training

The goal of the training of an ANN is to learn the weights and biases that allow us to make reasonable predictions about the data sets that interest us. As we discussed earlier, we use the back-propagation algorithm to perform the training. Practically, this means that our task at this point is to set up CAS so that it can run the training algorithm. Setting up the training algorithm mostly involves choosing the training hyper-parameters and letting the `deepLearn.dlTrain` action do the work for us.

The training hyper-parameters that we use are as follows:

The Optimizer

The methodology that we use to minimize the loss is the adaptive moment estimation or Adam optimizer. The Adam optimizer is a derivative of the gradient descent method (sometimes called the steepest descent algorithm).

The Learning Rate

The learning rate drives how quickly the gradient updates follow the direction of the gradient. Practically, if the learning rate is too small (for example, 10^{-10}), the training converges too slowly and consequently takes a long time. By the same token, if the learning rate is too big (for example, 10), the optimizer is likely to miss the minima and fail to converge. Choosing the "right" value is an art. A good strategy is to start at 10^{-2} and see how well we do.

The Mini Batch Size

This hyper-parameter governs how many observations we submit per iteration. The main advantage is that with a large enough mini batch size, we compute the gradient over several observations and introduce enough noise to avoid saddle points that would prevent us from converging. Another advantage of the mini batch size is the computational implication. Specifically, if we had only one core, then one observation at a time would work just fine computationally (not numerically). However, with many CPU cores and especially GPUs, you want to submit multiple observations at the same time. Here again, the selection of the hyper-parameter is mostly an art, and starting at 100 is a good strategy.

GPU

By setting a value of 1, we request to use the available GPUs during the training. If you have a GPU, you should always use it unless you suspect a computational issue (in other words, a bug) or you know for a fact that the GPU won't work (if you use a custom loss function that cannot run on the GPU, for example).

The Seed

Adam is a stochastic optimizer, so the seed is used to initialize a pseudo-random number generator.

The Number of Epochs

In machine learning, an epoch is a pass through the entire training set. A higher number of epochs provides the training algorithm with more iterations to reach a satisfactory loss.

The Validation Frequency

This hyper-parameter governs how often we compute the loss function over the validation data set. This is given in number of iterations.

The loss function that we need to minimize is not a training hyper-parameter, since we specified it as a model hyper-parameter earlier (for a logistic regression, we rely on the cross-entropy).

Practically speaking, the code to invoke the `deepLearn.dlTrain` action is rather straightforward:

Program 2.9: Invoking the deepLearn.dlTrain Action

```
deepLearn.dlTrain result = results /
    table        = training_table
    model        = {name = model_name}
    modelWeights = {name = model_name || "_param", replace = TRUE}
    inputs       = continuous_inputs
    target       = target
    nominal      = categorical_inputs
    seed         = 144
    gpu          = 1
    validFreq    = 10
    validTable   = validation_table
    optimizer    = {
            algorithm     = {method = "Adam" learningRate = 0.01}
            maxEpochs     = 500,
            miniBatchSize = 50
    };
run;

saveresult results['OptIterHistory'] replace
        dataout = dlTrain_iterations;
run;
```

You will notice that in addition to specifying the hyper-parameters that we mentioned, we also save the results into a CAS table called `results` and then extract one column that we write to a SAS 9.4 table, `WORK.dlTrain_iterations`. We will see later why this is good practice.

One additional point that we should make before running is that in CAS, one separates the model (`model = "iris_logistic_regression"`) from the model weights and biases (`modelWeights = "iris_logistic_regression_param"`). This separation is to be expected, because you define the model once and you train it many times.

We pointed out earlier that the inputs (and outputs) are specified during training as well, so let's have a look at them. We have two types of inputs and outputs:

Continuous
   Inputs or outputs with continuous values over an interval of real values (for example, the sepal and petal lengths).

Categorical
   Inputs or outputs with (few) discrete values (for example, the species of irises).

For the training, we have to distinguish the continuous inputs versus the categorical inputs, but not the outputs (since we clearly indicated that we were performing a logistic regression):

```
continuous_inputs  = {"PetalLength", "PetalWidth", "SepalLength",
                       "SepalWidth"};
categorical_inputs = { };
target             = "Species";
```

In this particular case, since we have no categorical inputs, we pass an empty list, represented by { }.

Running the training with the hyper-parameters that we have set should take a couple of seconds with GPUs (but not much longer on CPUs, since this is a relatively simple network):

```
154         deepLearn.dlTrain result = results /
155            table         = training_table
156            model         = {name = model_name}
157            modelWeights  = {name = model_name || "_param",
                                  replace = TRUE}
158            inputs        = continuous_inputs
159            target        = target
160            nominal       = categorical_inputs
161            seed          = 144
162            gpu           = 1
163            validFreq     = 10
164            validTable    = validation_table
165            optimizer     = {
166                algorithm     = {method = "Adam"
                                    learningRate = 0.01}
167                logLevel      = 1,
168                maxEpochs     = 500,
169                miniBatchSize = 50
170            };
171         run;
NOTE:  Synchronous SGD is starting.
NOTE:  The total number of parameters is 16495.
NOTE:  The approximate memory cost is 1.00 MB.
NOTE:  Initializing each layer cost     1.24 (s).
NOTE:  The total number of threads on each worker is 64.
NOTE:  The total number of minibatch size per thread on each worker
       is 50.
NOTE:  The maximum number of minibatch size across all workers for
       the synchronous mode is 3200.
NOTE:  The optimization reached the maximum number of epochs.
NOTE:  The total time is     1.88 (s).
```

If you have an NVIDIA GPU, you can check that the GPU did some work (it might be hard to catch the GPU in the act since the training is so quick). For example, in LINUX you would enter the following command:

```
[dl4na@fsnlax05 ~]$ nvidia-smi
+-----------------------------------------------------------------------------+
| NVIDIA-SMI 375.26                 Driver Version: 375.26                     |
|-------------------------------+----------------------+----------------------+
| GPU  Name        Persistence-M| Bus-Id        Disp.A | Volatile Uncorr. ECC |
| Fan  Temp  Perf  Pwr:Usage/Cap|         Memory-Usage | GPU-Util  Compute M. |
|===============================+======================+======================|
|   0  Tesla K80           Off  | 0000:05:00.0     Off |                    0 |
| N/A   36C    P0    63W / 149W |     92MiB / 11439MiB |     52%      Default |
+-------------------------------+----------------------+----------------------+
|   1  Tesla K80           Off  | 0000:06:00.0     Off |                    0 |
| N/A   30C    P0    73W / 149W |     92MiB / 11439MiB |     26%      Default |
+-------------------------------+----------------------+----------------------+
```

In the preceding example, we have two NVIDIA K40 GPUs (labeled as a K80 card).

Figure 2.13 shows the training and the validation loss.

Figure 2.13: Training and Validation Loss for IRIS Classification

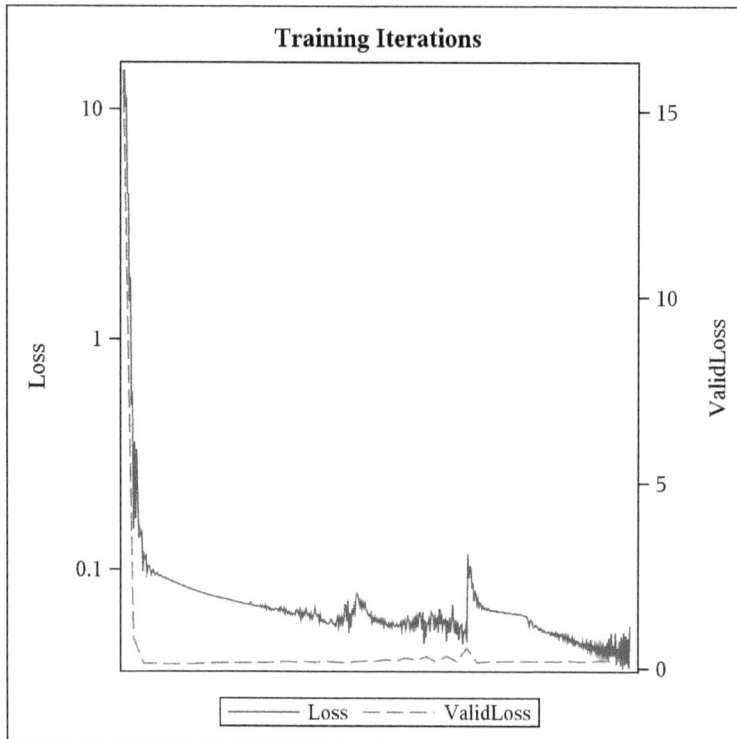

The shape of the loss functions over the iterations is almost textbook perfect. We want to see an L-curve like the ones that you see on that figure. That shape indicates that the algorithm is learning, quite efficiently in this case. Please note that the Y axis on the left is for the loss of the training set, whereas the Y axis on the right is for the loss of the validation set. Both Y axes use a logarithmic scale.

Sometimes the Y axis of Figure 2.13 is reversed and you get a learning curve. This observation contradicts the common belief about learning curves: steep learning curves are your friends in machine learning (LeCun, 2004).

Training is one thing, but scoring is where things start to matter. How well can our network classify observations that it has never seen? Let's have a look.

Inference

Inference in ML is narrower than what is traditionally called *Statistical Inference*. In ML inference, we are simply interested in making a prediction, which in our case translates to feeding an input set to the network in Figure 2.13 and collecting the probabilities for each class.

With CAS, inference is called scoring, and the action to perform scoring on a previously trained model is `deepLearn.dlScore`:

```
deepLearn.dlScore /
    table        = testing_table
    model        = model_name
    initWeights  = {name = model_name || "_param"}
    copyVars     = {'Species'}
    gpu          = 1
    casout       = {name = scoring_table, replace = TRUE};
```

We specify the model along with the weights and biases. In addition, we specify the table with the testing observations (`table = "iris_test"`). You'll notice that we explicitly carry over the species to the scoring table to be able to tell which is which when we look at the observations after the inference.

The results of running the inference on our small testing set are shown in Figure 2.14.

Figure 2.14: Scoring for Iris Classification

Results from deepLearn.dlScore

Score Information for IRIS_TEST	
Number of Observations Read	20
Number of Observations Used	20
Misclassification Error (%)	0
Loss Error	0.022546

Bull's eye! We have a perfect score: all 20 testing samples have been correctly classified. Let's have a closer look at the results of the scoring:

Obs	Species	_DL_P0_	_DL_P1_	_DL_P2_	_DL_PredName_
1	Versicolor	0.999880	0.000120	0.000000	Versicolor
2	Virginica	0.243513	0.756487	0.000000	Virginica
3	Setosa	0.000000	0.000000	1.000000	Setosa
4	Versicolor	0.999258	0.000740	0.000003	Versicolor
5	Versicolor	0.907960	0.092040	0.000000	Versicolor
6	Virginica	0.000875	0.999125	0.000000	Virginica
7	Setosa	0.000000	0.000000	1.000000	Setosa
8	Versicolor	0.999417	0.000583	0.000000	Versicolor
9	Virginica	0.003128	0.996872	0.000000	Virginica
10	Virginica	0.295860	0.704140	0.000000	Virginica
11	Versicolor	1.000000	0.000000	0.000000	Versicolor
12	Versicolor	0.999594	0.000000	0.000406	Versicolor
13	Versicolor	0.999990	0.000000	0.000010	Versicolor
14	Setosa	0.000000	0.000000	1.000000	Setosa
15	Setosa	0.000000	0.000000	1.000000	Setosa
16	Setosa	0.000000	0.000000	1.000000	Setosa
17	Virginica	0.000122	0.999878	0.000000	Virginica
18	Versicolor	0.999978	0.000020	0.000002	Versicolor
19	Versicolor	0.999982	0.000000	0.000018	Versicolor
20	Virginica	0.043318	0.956682	0.000000	Virginica

In the preceding table, you can see that the network is quite sure of itself. Only two of the prediction probabilities are below 0.9 (_DL_P0_ is the probability of the first class computed by the first output neuron, _DL_P1_ is for the second neuron, and _DL_P2_ for the last one).

We can run the scoring on the entire set of 150 observations and get the scoring results listed in Figure 2.15.

Figure 2.15: Total Scoring for IRIS Classification

Results from deepLearn.dlScore

Score Information for IRIS_EVERYTHING	
Number of Observations Read	150
Number of Observations Used	150
Misclassification Error (%)	0.666667
Loss Error	0.031401

In this case, we got one misclassification:

Obs	Species	_DL_P0_	_DL_P1_	_DL_P2_	_DL_PredName_
58	Versicolor	0.038895864	0.961104095	2.77E-13	Virginica

This is not a big surprise if you examine Figure 2.9 and notice that several cases of one Versicolor observation are sitting right on top of one Virginica observation. Without an alternate feature representation and probably more data, it would be difficult to increase the accuracy over the entire set.

This concludes our introduction to logistic regression with DL. In the next chapter, we will concentrate on regressions, beginning with a simple regression problem with the CARS data set.

Conclusion

We started this chapter by looking at the history of machine learning and deep learning. We saw that the DL networks that we use today are multilayer perceptron networks that we train using the back-propagation algorithm. We also saw that for decades, the main focus of artificial intelligence and DL had been classification.

We discussed the importance of machine learning as opposed to programming when it comes to learning from data.

We then looked at some data sets that come with any SAS installations to find suitable candidates for ML projects for classifications.

After a quick introduction to SAS Viya and CAS, we saw how easily and naturally we could program a DL algorithm in SAS by using SAS 9.4 and SAS Viya. In a few lines of code, we rapidly created a fully connected deep neural network with two hidden layers.

After quickly training our network on GPUs, we noticed that we had a perfect score on the classification of our test set. We could achieve this result with little to no domain knowledge.

In the next chapter, we will look deeper at ML with CAS using regressions as an example.

Chapter 3: Regressions

In Chapter 2, we started to get acquainted with the deep learning (DL) capabilities of SAS Viya by using the CAS server from SAS Studio. We were quite successful at classifying data sets using deep learning techniques, so it is time to see whether we can build on that success and learn how to solve numerical problems by performing regressions using a deep neural network (DNN).

A Brief History of Regressions

In this section, we give a very brief introduction to regressions and their history. If you're already familiar with regressions, please feel free to jump to the next section.

A regression analysis enables us to measure and analyze the relationships between two or more variables. More specifically, the problem that occupies our minds in this chapter is prediction: Based on the values observed in a training data set, how can we predict as accurately as possible the values of a dependent variable based on the values of the independent variable? By convention, we label the dependent variable Y and the independent variable X. In machine learning (ML), the variables denoted by X are usually called features.

A regression is similar to the classification that we investigated in the previous chapter (classification is also called logistic regression). The difference is that we are no longer interested in the discrete classes of the dependent variable, but in its actual value. In addition to this primary difference, as we will see shortly, there are some key differences in the methodology that we use to determine a satisfactory approximation of Y, the dependent variable.

Historically, regressions have been classified into linear and nonlinear regressions. In linear regressions (Figure 3.1), the estimator is a line. In nonlinear regressions (Figure 3.2), the estimator is something else—typically, a curve of a higher order than the line (for example, the polynomial curve of the 4th order in Figure 3.2).

Figure 3.1: A Linear Regression

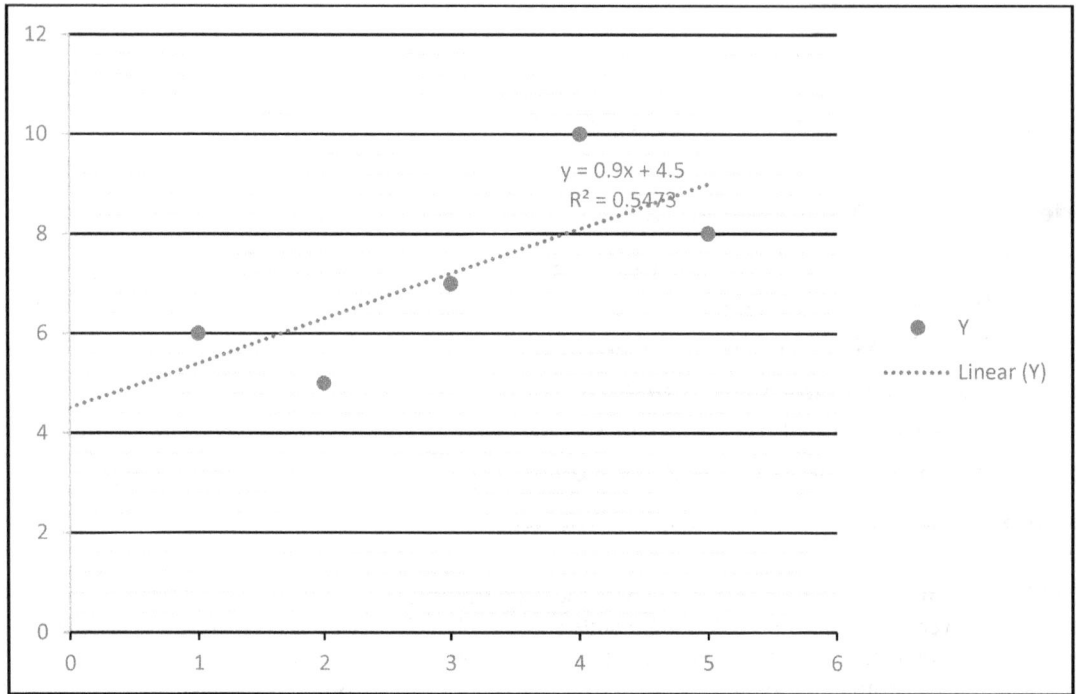

$y = 0.9x + 4.5$
$R^2 = 0.5473$

Figure 3.2: A Nonlinear Regression

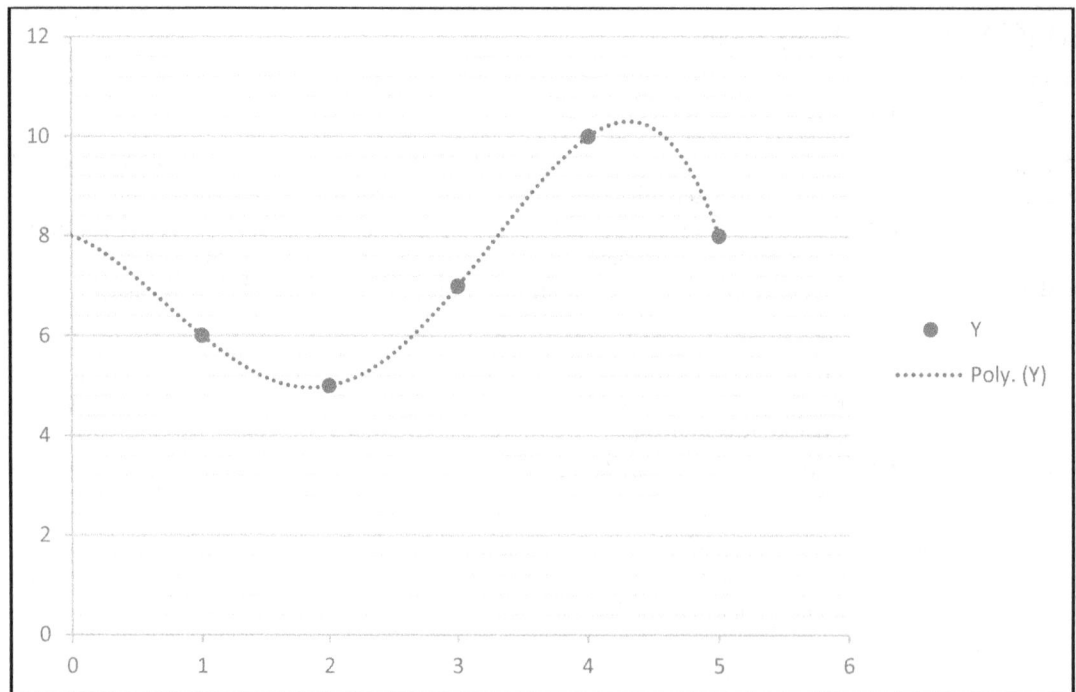

The exact date of the first use of regressions in mathematics is unknown, but it seems that the precursor was in the work of Roger Joseph Boscovitch, who geometrically calculated the coefficients of a linear regression around 1755. Shortly after, the marquis Pierre Simon de LaPlace, whom we will encounter again when we talk about Monte Carlo simulations, came up with an algebraic method to compute a linear regression (Howarth 2017) using Boscovitch's algorithm.

The least squares method that is widely used today for linear regressions was born in a halo of controversy because Carl Friedrich Gauss and Adrien Marie Legendre independently found the least squares method. Legendre first published a description in 1805 (Legendre 1805), but when Gauss learned about Legendre's publication, he pointed out that he already knew the least squares method. Gauss claimed that he didn't publish his invention because it was "so trivial." Gauss was able to prove his claim because he had previously published the results of cosmic calculations that could only have been obtained using the least squares method. Later, in 1809, Gauss published a more detailed and more formal study of the least squares method (Stigler 1981).

We would have to wait a few years for the word "regression" to appear in the context of a linear regression. In 1886, Francis Galton published his paper about human heredity, in which he predicted the heights of the children based on the height of the parents: *Regression Towards Mediocrity in Hereditary Stature* (Galton 1886). In that paper, Galton points out the *regression to the mean* of the tallness of offspring: if the parents are abnormally tall, the children tend to be of a height closer to the mean of the population.

Nonlinear regressions were introduced in the 1920s with the work of Ronald Fisher (1925) and others, but were mostly theoretical until the 1970s with the advent of computers and software systems such as SAS.

Based on our study of artificial neural networks (ANN) in the previous chapters, we already know that the ANN won't easily give out its secrets. This means that in most cases, you won't know at first glance what model the ANN approximates: A linear regression? A nonlinear regression? What type of nonlinear regression? A polynomial? Of what order? We will see shortly that despite not knowing the answers to those questions, we can see clues that point to one model or another. In other words, an ANN is not a complete black box.

As we mentioned repeatedly in the previous chapter, looking at the data during each stage of a ML project is a good practice. So, let's start by looking at the training data set that we will use for regressions.

All about the Data (Reprise)

The CARS Data Set

The SASHELP.cars data set contains 428 observations of car classes and measurements. Our plan is to use the CARS data set for a simple regression.

Figure 3.3 shows the basic information that we extract after loading the data set into CAS. There are no surprises here.

Figure 3.3: Results of PROC CASUTIL for the CARS Data Set

Caslib Information	
Library	CASUSER(dl4na)
Source Type	PATH
Description	Personal File System Caslib
Path	/home/dl4na/casuser/
Session local	No
Active	Yes
Personal	Yes
Hidden	No
Transient	Yes

Table Information for Caslib CASUSER(dl4na)										
Table Name	Number of Rows	Number of Columns	Indexed Columns	NLS encoding	Created	Last Modified	Promoted Table	Repeated Table	View
CARS	262	15	0	utf-8	19Mar2018: 12:25:14	19Mar2018: 12:25:14	No	No	No

Table Information for Caslib CASUSER(dl4na)	
Table Name	Compressed
CARS	No

We then look at a scatter plot of the variables that we intend to use for regression. We would like to predict the miles per gallon (MPG) based on the manufacturer's suggested retail price (MSRP) because we suspect that sticker shock correlates to pump shock. In case you're not familiar with this colloquialism, Merriam-Webster defines sticker shock as "astonishment and dismay experienced on being informed of a product's unexpectedly high price." So we would define pump shock as "astonishment and dismay experienced on being informed of a car's unexpectedly poor mileage."

In Figure 3.4, we plotted the MPG for the city and the highway. Right away we see some outliers. For example, an MPG of approximately 60 looks suspicious. A careful look at the data reveals that it is for hybrid cars:

Make	Model	Type	MSRP	MPG_City	MPG_Highway
Honda	Civic Hybrid	Hybrid	$20,140	46	51
Honda	Insight 2dr	Hybrid	$19,110	60	66
Toyota	Prius 4dr	Hybrid	$20,510	59	51

Note that there hasn't been a significant drop in car prices while you were reading this book: those prices are from 2004.

So this data suggests that we should probably perform this regression per class (for example, sedan versus SUV versus hybrid).

Figure 3.4: Results of PROC SGPLOT for the CARS Data Set

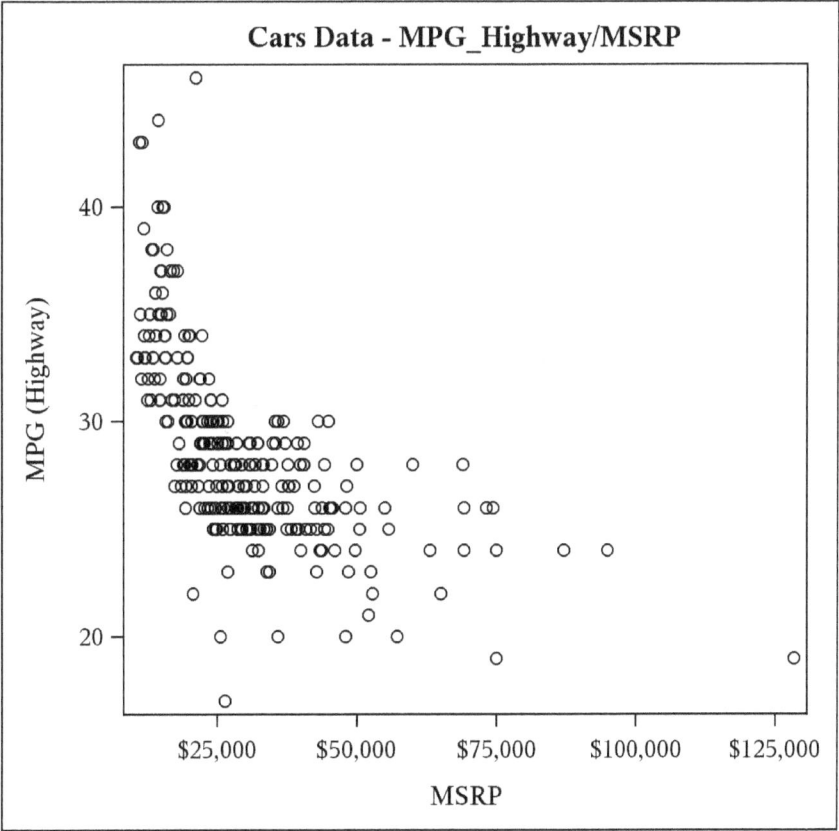

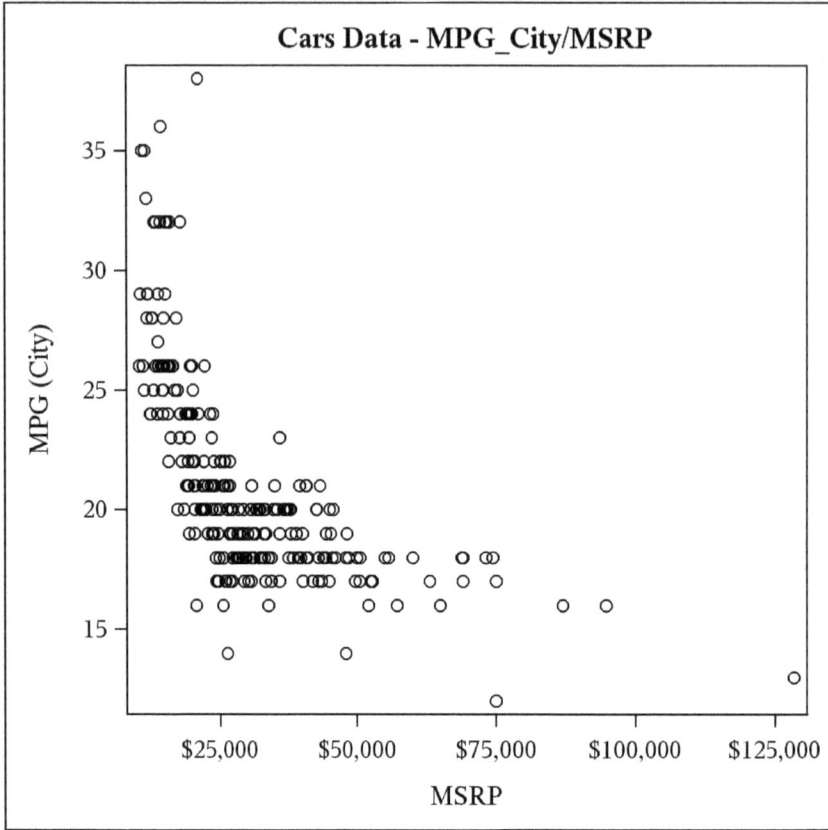

Now that we have data for a regression, let's build a model for it.

A Simple Regression

As we did in the previous chapter when we looked at classifications, the first thing that we need to do is create a CAS session:

Program 3.1: Opening a CAS session

```
options cashost='fsnlax05' casport=5570;
cas mysession;

libname DL_LIB cas sessref = mysession;
```

Once we have a CAS session, we shuffle our data and create three data sets: one for training, one for validation, and one for testing:

Program 3.2: Creating Data Sets for Training, Validation, and Testing

```
data cars;
  set SASHELP.cars;
  where type = 'Sedan';
  sortval = rand('UNIFORM');
run;

proc sort data=cars out=cars(drop=sortval);
  by sortval;
run;

/* Load the training data set into CAS */
data DL_LIB.cars_train;
   set cars(obs=200);
run;

/* Load the validation data set into CAS */
data DL_LIB.cars_validation;
   set cars(firstobs=201 obs=230);
run;

/* Load the testing data set into CAS */
data DL_LIB.cars_test;
   set cars(firstobs=231);
run;
```

At this point, in CAS, we have the situation summarized in Figure 3.5.

Figure 3.5: The Training, Validation, and Test Data Sets for CARS.

Caslib Information	
Library	CASUSER(dl4na)
Source Type	PATH
Description	Personal File System Caslib
Path	/home/dl4na/casuser/
Session local	No
Active	Yes
Personal	Yes
Hidden	No
Transient	Yes

Table Information for Caslib CASUSER(dl4na)							
Table Name	Number of Rows	Number of Columns	Indexed Columns	NLS encoding	Created	Last Modified	Promoted Table
CARS_TRAIN	200	15	0	utf-8	19Mar2018:12:34:55	19Mar2018:12:34:55	No
CARS_VALIDATION	30	15	0	utf-8	19Mar2018:12:34:55	19Mar2018:12:34:55	No
CARS_TEST	32	15	0	utf-8	19Mar2018:12:34:55	19Mar2018:12:34:55	No

Table Information for Caslib CASUSER(dl4na)			
Table Name	Repeated Table	View	Compressed
CARS_TRAIN	No	No	No
CARS_VALIDATION	No	No	No
CARS_TEST	No	No	No

We now have data, so we need a model. As we did in the previous chapter, we define a few CAS variables (notice that we use 10 neurons, the exact same number as we did for classification):

Program 3.3: Defining Variables

```
proc cas;

/* First, we build the model */
model_name                   = "cars_regression";
training_table               = "cars_train";
validation_table             = "cars_validation";
testing_table                = "cars_test";
scoring_table                = "cars_score";
number_of_hidden_layers      = 10;
number_of_neurons_per_layer  = 10;
```

We then build the model and add an input layer (still no changes compared to the classification network):

Program 3.4: Building the Model and Adding Layers

```
deepLearn.buildModel/
  model = {name = model_name, replace = 1}
  type  = "DNN";

/* First layer is the input layer */
deepLearn.addLayer result = add_layer_results/
  layer = {type = "INPUT"}
  name  = "input"
  model = {name=model_name};

/* Then we add the hidden layers */
previous_layer_name = "input";
do i = 1 to number_of_hidden_layers;
  new_layer_name = "hidden_" || i;

  deepLearn.addLayer result = add_layer_results/
    layer     = {type = "FC", act = "RELU",
                 n = number_of_neurons_per_layer}
    name      = new_layer_name
    srcLayers = {previous_layer_name}
    model     = {name = model_name};

    previous_layer_name = new_layer_name;
end;
```

For the output layer, things are a little different.

The first change has to do with the output of the network. We're no longer interested in the probability of the classes, but rather in one continuous value in $\mathbb{R}$. To get that behavior, we simply specify the IDENTITY as the activation function. The output layer computes the linear combination of the outputs of the neurons of the last hidden layer.

The second change has to do with the error (or loss) function that we use during training. Since we have only one value, we can rely on the mean squared error (MSE):

$$\frac{1}{n}\sum_{i=1}^{n}(\widehat{Y_i} - Y_i)^2$$

where n is the number of observations, 1 in this case.

Using the MSE is achieved by specifying `error = "NORMAL"`:

Program 3.5: Adding the Output Layer

```
/* Last layer is the output layer */
deepLearn.addLayer result = add_layer_results /
  layer      = {type = "OUTPUT", act = "IDENTITY", error = "NORMAL"}
  name       = "output"
  srcLayers  = {previous_layer_name}
  model      = {name = model_name};
run;
```

As you will see shortly, when we print and plot the result, specifying `error = "NORMAL"` results in using half of the MSE as a loss function.

At this point, our regression model is complete, so we can train it. As you probably recall from last chapter and have undoubtedly noticed in the preceding code, we haven't specified any inputs or outputs so far. That's because specifying inputs and outputs is part of the training, not of the model definition. So, let's specify the inputs and outputs:

```
continuous_inputs   = {"MSRP"};
categorical_inputs = { };
target              = "MPG_city";
```

There are no categorical inputs in this case (partly because we are only training the estimator for one class), so we use an empty list: { }. The invocation of the training loop is done via the `deepLearn.dlTrain` action:

Program 3.6: Invoking the Training Loop with the deepLearn.dlTrain Action

```
deepLearn.dlTrain result = results /
   table        = training_table
   model        = {name = model_name}
   modelWeights = {name = model_name || "_param", replace = TRUE}
   inputs       = continuous_inputs
   target       = target
   nominal      = categorical_inputs
   seed         = 144
   gpu          = 1
   validFreq    = 10
   validTable   = validation_table
   optimizer    = {
         algorithm     = {method = "ADAM" learningRate = 0.01}
         maxEpochs     = 1,
         miniBatchSize = 140
   };
```

We have increased the mini-batch size a bit, because we have more observations than we had for the training of the IRIS classifier. We've set `maxEpochs = 1` to illustrate what the ANN does with little training, namely one run through the training data.

Figure 3.6: Predictions after 1 Epoch

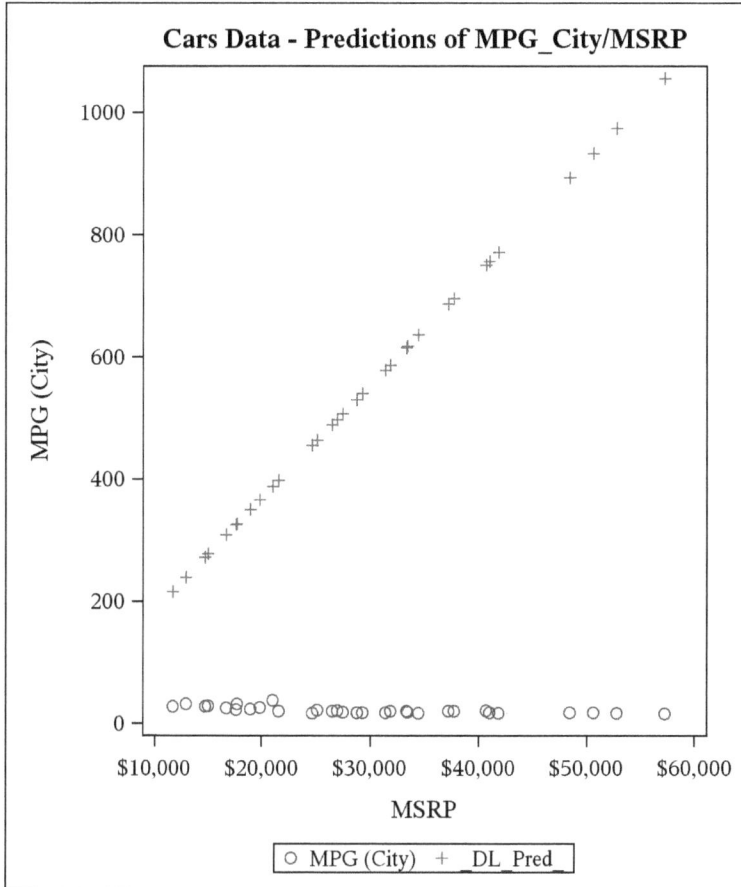

In Figure 3.6, we show the test set true value for MPG_CITY with the predicted values (under the label _DL_PRED) after 1 epoch of training. As you can see, the network predicts a straight line, essentially performing a (poor) linear regression. You can see in Figure 3.7 that as the number of training iterations increases to 1,000, the quality of the predictions improves. You might also notice in Figure 3.8 that the model is moving away from a linear regression.

Figure 3.7: Loss for 1,000 Epochs

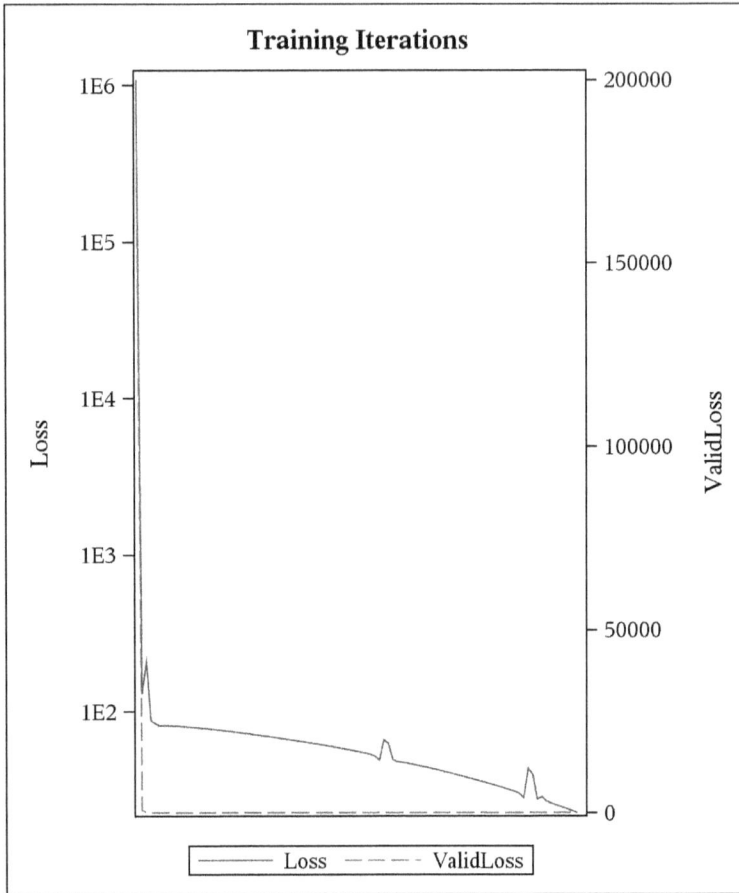

Figure 3.8: Predictions after 1,000 Epochs

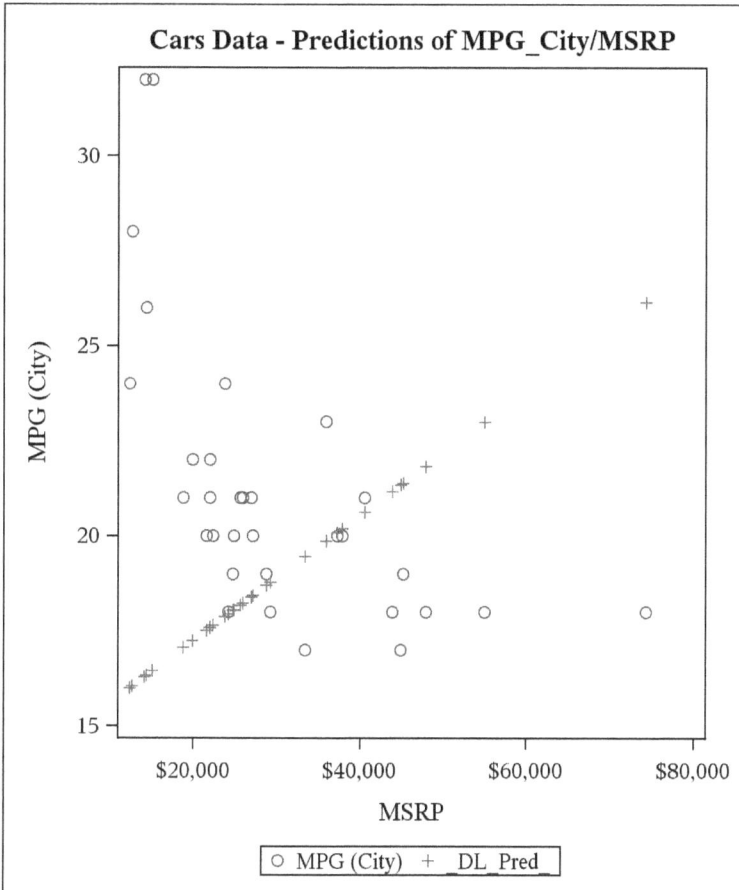

If we increase the number of training loops so that the network can converge close to its best possible answer, the model is no longer linear at all. This is obvious in Figure 3.9. It is also self-evident at this point that sticker shock readily translates into pump shock.

At the end of the training with 5,000 epochs and after the scoring of our test set, we can observe the following values for the loss:

Training	Validation	Scoring
3.03	2.48	4.14

Figure 3.9: Predictions after 10,000 Epochs

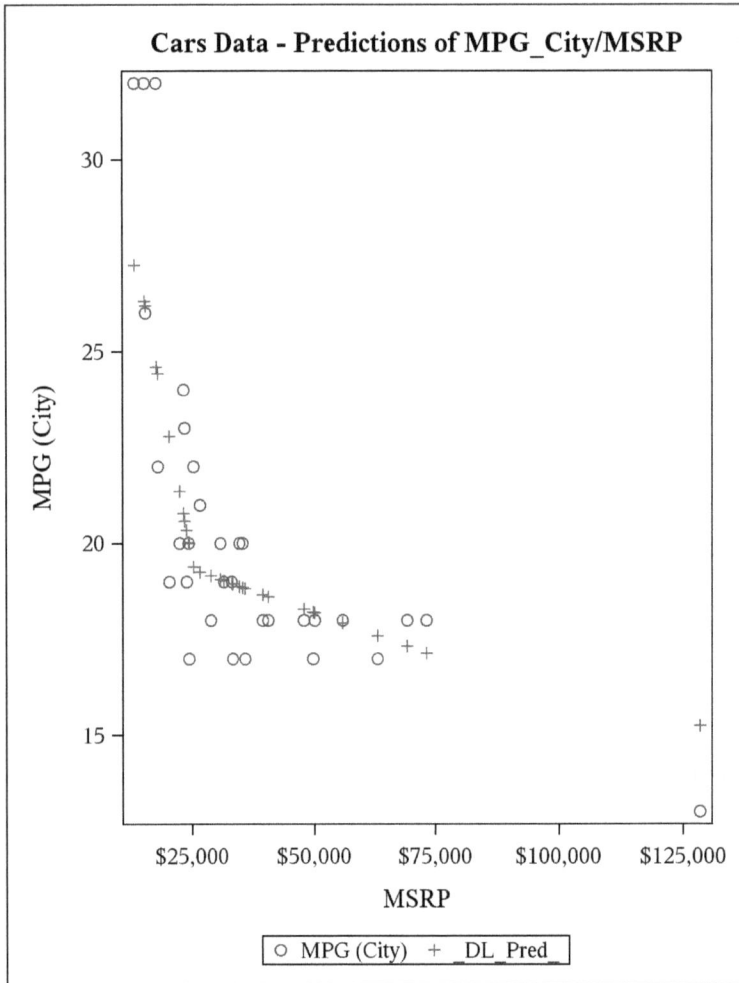

If you look at where we plot the loss function for 10,000 iterations, you will notice that after 6,000 iterations, we essentially lose our time: the algorithm is not learning much anymore.

Figure 3.10: Loss for 10,000 Iterations

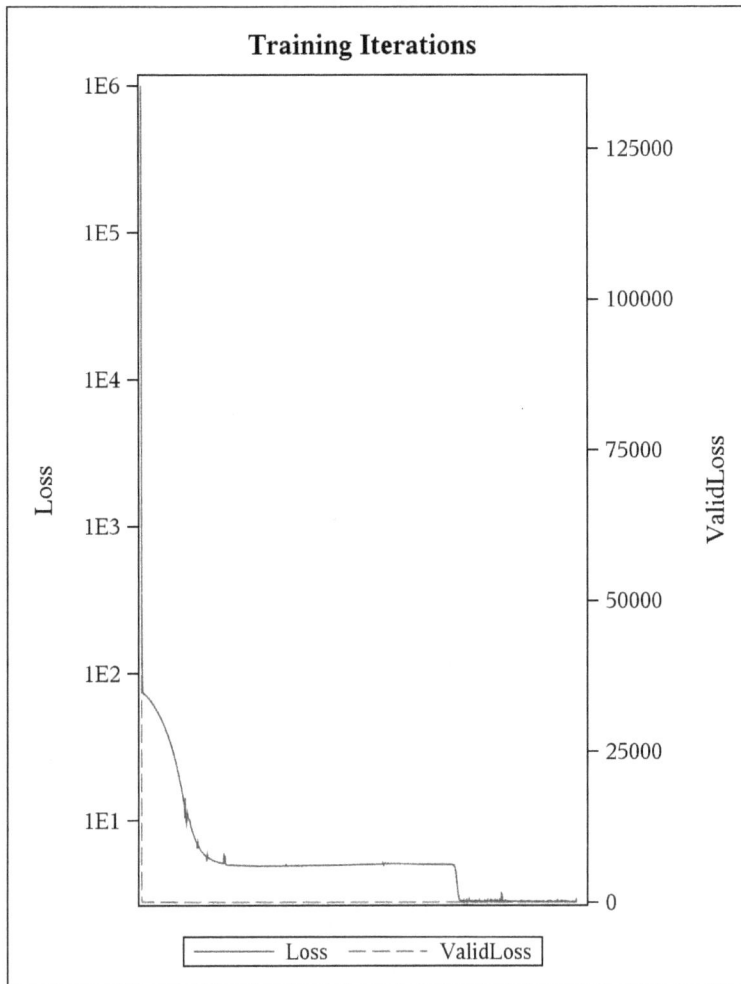

This is an example where early stopping would have made sense. Alternatively, more data might have helped, if there was more to learn.

You might have to try multiple runs in order to reproduce similar results because of the stochastic nature of the training, the selection of the training/validation/testing triplet, and the small size of our training data set (a few changes in the training data have a significant impact on the training).

As you can see in the preceding code, the classification and the regression model are identical except for a few hyper-parameters:

- the activation function of the last layer. For the regression, we used a simple linear combination rather than the softmax function.
- the error function of the last layer. For the regression, we used the MSE rather than the cross-entropy function.

Other than those changes, we used the same network architecture and the same training algorithm. This uniqueness of the algorithm is a great strength of DL that enables us to leverage custom hardware in an economical fashion to significantly speed up our numerical applications. We will revisit this discussion in more depth in Chapter 9.

Note that to have a truly unique algorithm that handles classifications and regressions, you might be tempted to handle a classification like a regression with one output rather than softmax outputs. This would probably work in most cases, but would come with a couple of drawbacks:

1. The training will not converge as quickly in some cases. This is because the gradient of an output that should be 1 can be very small at the beginning, so it could take a long time (many iterations) for the output to get close to 1.

2. The interpretation of the output is less palatable than the probability of the class.

Essentially, if you were to use a regression to perform a classification, you would be depriving your network of the information that the output should be a probability, so the training would have to work harder to give you satisfactory results.

Now that we have a good understanding of classification and regression with a deep neural network based on data, we can focus our attention to the approximation of a continuous function. In this case, we typically don't have the data, so we must generate that data.

The Universal Approximation Theorem

In layman's terms, the *universal approximation theorem* states that a multilayer perceptron (MLP) network with a single hidden layer containing a finite number of neurons can approximate a continuous function on a compact subset of $\mathbb{R}^n$. The universal approximation theorem is often called Cybenko's theorem because George Cybenko proved the theorem in 1989 for the sigmoid activation function (Cybenko 1989). A couple of years later, Kurt Hornik proved that the theorem was applicable for any activation function, not just the sigmoid function (Hornik 1991).

The universal approximation theorem is a pertinent theorem for our purpose because it provides a theoretical framework for the approximation of continuous functions using deep neural networks. And as we will see in Chapter 4, by relying on many-task computing to compute analytics, we define a function for each SAS program. Hornik's proof is relevant to us, because we will rely on the ReLU activation function instead of the sigmoid activation function.

It is important to point out that the theorem doesn't provide a methodology to define the architecture of the network (the required number of neurons is finite, but what is it?). Neither does it provide a methodology to train the network (for example, should we use Adam? Statistical gradient descent? What learning rate?). The theorem simply tells us that the network exists, so at least we have a shot!

In practice, we will see that following a few basic rules enables us to approximate the analytics that we consider in this book with a satisfactory level of accuracy.

Universal Approximation Framework

In this section, we build a framework to approximate a continuous $\mathbb{R}^n \to \mathbb{R}$ function. For our purpose, we define a framework as reusable code. As you will see shortly, it would be easy to extend the framework to handle $\mathbb{R}^n \to \mathbb{R}^m$ functions. We stick to $\mathbb{R}^n \to \mathbb{R}$ for the sake of simplicity in this introductory chapter.

So why do we need a framework? A DL framework is useful for a couple of reasons:

- We need to separate the learning or training from the inference.
  So far, our examples have been quite simple, so the training would take a few seconds. This is about to change. The ratio between training and inference is usually orders of magnitude (for example, it takes 15 minutes to train a network that runs inference in less than a second). We hide the complexity of saving and restoring the DNN in a couple of functions of the framework.
- Our training and inference are universal, so we want to reuse as much code as possible.
  Give or take a few hyper-parameters, we use the same algorithm for the rest of book. So we might as well cast that algorithm in stone and make it explicit.

For the definition of our framework, we could either rely on SAS macros or the CASL language that is supported by the CAS procedure (PROC CAS). Since all our training and inference is done inside CAS, we choose the latter. We will also see shortly that PROC CAS has a few key advantages over SAS macros.

We separate our DL framework into two files:

- `model_io.sas`
  This module contains the functions that deal with saving and restoring a DL model.
- `ua.sas`
  This module contains our implementation of the universal approximation theorem.

Those two files are available in the same folder as the code for this chapter.

Let's examine our DL framework bottom to top, starting with the input and output functions in `model_io.sas`.

Before writing some code, we should point out that it takes three tables to save a DL model in CAS:

1. One table contains the definition of the model (how many layers, how many neurons, the activation function, and so on).
   By convention, we call the definition table `<model-name>` (for example, `function`). On disk, the table is always saved with the extension `sashdat` (for example, `function.sashdat`). Note that `sashdat` is the extension used by CAS; it is not a `sas7bdat` 9.4 data set.
2. One table contains the definition of the parameters (weights and biases) of the neurons.
   By convention, we call the parameter table `<model-name>_weights` (for example, `function_weights`).
3. One table contains the attributes of the weights.
   In CAS, the data type of a column is an attribute of the column. Since the table attributes are stored in a different table, we need to specify those attributes every time we save and restore a model.
   By convention, we call the attribute table `<model-name>_weights_attr` (for example, `function_weights_attr`).

In this book, we save all tables in our default personal caslib, `CASUSER(dl4na)`.

To easily manage the three tables, we provide the following functions:

Program 3.7: Functions for Managing Tables

```
function get_param_table_name(model_name);
  return model_name || "_weights";
end func; /* get_param_table_name */

function get_attr_param_table_name(model_name);
  return model_name || "_weights_attr";
end func; /* get_attr_param_table_name */
```

Please note the usage of || as a concatenating operator. Other than that, the syntax of the definition is self-explanatory. As you might have guessed by looking at the code, CASL is a dynamically typed language. The type of the arguments is inferred based on the context; they are not just text as in a macro language. An in-depth presentation of the CASL language can be found in the SAS documentation (A Guide to the SAS 9.4 and SAS Viya Programming Documentation).

We need a function to store (and replace) a table so that we have a convenient way to save and replace the three tables that make up a DL model:

Program 3.8: Function for Storing and Replacing a Table

```
function save_and_replace_table(table_name,target_library_name);
  table.save/
    table   = table_name           /* source table name        */
    caslib  = target_library_name     /* target caslib name        */
    name    = table_name              /* target table, contains
                                        sub-folders, if any      */
    replace = TRUE;
end func;  /* save_and_replace_table */
```

In the `save_and_replace_table` function, we rely on the `table` action set, which contains actions such as `save` and `load` that we can leverage in the `load_table` function:

Program 3.9: Function for Loading a Table

```
function load_table(cas_library_name, table_name, attr_table_name);
  table.loadTable/
    caslib = cas_library_name
    casOut = table_name
    path   = table_name || ".sashdat";

  /* isString is not missing and not 0/NULL  */
  if isString(attr_table_name) then do;
    /* Apply attribute table to parameter table. The (model)
       parameter table contains the weights and biases */
    table.attribute /
      table  = attr_table_name
      caslib = cas_library_name
      name   = table_name
      task   = "ADD";
  end do;
end func;  /* load table */
```

The test `if isString(attr_table_name) then do` enables us to conveniently load the attribute of a table with the table itself.

Armed with these basic input and output functions, we can now save a DL model to disk:

Program 3.10: Function for Saving a DL Model to Disk

```
function save_model(model_name, cas_library_name);
  save_and_replace_table(model_name, cas_library_name);

  param_table_name      = get_param_table_name(model_name);
  attr_param_table_name = get_attr_param_table_name(model_name);

  save_and_replace_table(param_table_name, cas_library_name);

  /* extract the param_table attributes into a table
    (in the current caslib) and save that table */
  table.attribute /
    table  = attr_param_table_name
    caslib = cas_library_name
    name   = param_table_name
    task   = "CONVERT";
  save_and_replace_table(attr_param_table_name, cas_library_name);
end func;   /* save_model */
```

The only new code in the `save_model` function is the use of the `table.attribute` action to create the table with the attributes (for example, data types) of the model parameters (weights and biases).

The `load_model` function is now straightforward:

Program 3.11: Function for Loading a Model

```
function load_model(cas_library_name, model_name);
  param_table_name      = get_param_table_name(model_name);
  attr_param_table_name = get_attr_param_table_name(model_name);

  load_table(cas_library_name, model_name);
  load_table(cas_library_name, attr_param_table_name);
  load_table(cas_library_name, param_table_name,
             attr_param_table_name);
end func;   /* load_model */
```

Now that we have a set of input and output functions for our DL, we need a few functions for the training and inference. The body of those functions is defined in `ua.sas`. If you were to look at the code, you would see that it is akin to what we did earlier in this chapter (and the previous one). Here are the function signatures:

Program 3.12: Functions for Training and Inference

```
function define_model(model_name, number_of_hidden_layers);
function train(model_name, training_table, validation_table,
               inputs, target, nominals, max_epochs, use_gpu);
function inference(model_name, table_name,
                   scoring_table_name, copy_vars);
```

Most of the argument names are self-explanatory, but if in doubt, refer to `ua.sas`, which contains the headers with more explanations of the arguments.

In the `inference` function, the `table_name` is the table to score and the `scoring_table` is the result of the scoring. The `copy_vars` argument enables us to specify a list of the variables (columns) to copy from `table_name` to `scoring_table`.

We're almost done with our DL framework. We need to add one more helper function:

Program 3.13: Function for Setting Up Training Data

```
function setup_training_data(observations, training_percent);
  sampling.srs/
    table    = observations
    partInd = TRUE
    samppct = training_percent
    seed     = 1234
    output   = {
       casout   = {name = "temp_sampling", replace = TRUE},
       copyVars = "all"
    };

  training_table = {name = "temp_sampling",
                       where = "1 = _partind_"};
  validation_table = {name = "temp_sampling",
                        where = "0 = _partind_"};

  rvalue.training = training_table;
  rvalue.validation = validation_table;

  return rvalue;
end func; /* set_up_training_data */
```

The `setup_training_data` function takes a CAS table as input and splits it into a training set and a validation set based on the percentage passed in. We rely on the `sampling.srs` action, which partitions the input table into two partitions, one with the variable `_partind_` set to 0 and one with the `_partind_` variable set to 1.

The return value of the `setup_training_data` function is a structure that the calling function can use:

```
tables            = setup_training_data({name = training_table}, 80);
training_table    = tables.training;
validation_table = tables.validation;
```

Now that we have a DL framework, we can leverage it to learn a continuous function.

Approximation of a Continuous Function

In this section, we build a DL model for a $\mathbb{R}^1 \to \mathbb{R}$ function:

$$\sin(x) * pdf('normal', x, 0, 2)$$

More information on the pdf function can be found in the *SAS 9.4 Functions and CALL Routines: Reference* (SAS Institute, 2018). Figure 3.11 shows a plot of the function with x in [-10, 10]. The function is relatively complex to approximate, so it enables us to experience a more difficult training with our universal approximation framework. The function is nonrandom, but in Chapter 7, we will re-use our framework to approximate random functions.

The plot in Figure 3.11 is produced with the code in dl_load_function.sas, which in addition to plotting the curves produces a training data set of 10,000 observations:

Program 3.14: Generating Training Data

```
data INPUT.function_training;
  do x = -10 to 10 by 0.002;
    y = sin(x) * pdf('normal', x, 0, 2);
    output;
  end;
run;

proc cas;
  table.save/
    table = "function_training"
    name  = "function_training" /* The name is required */
    replace = TRUE;
quit;
```

Figure 3.11: A Continuous Function for Deep Learning

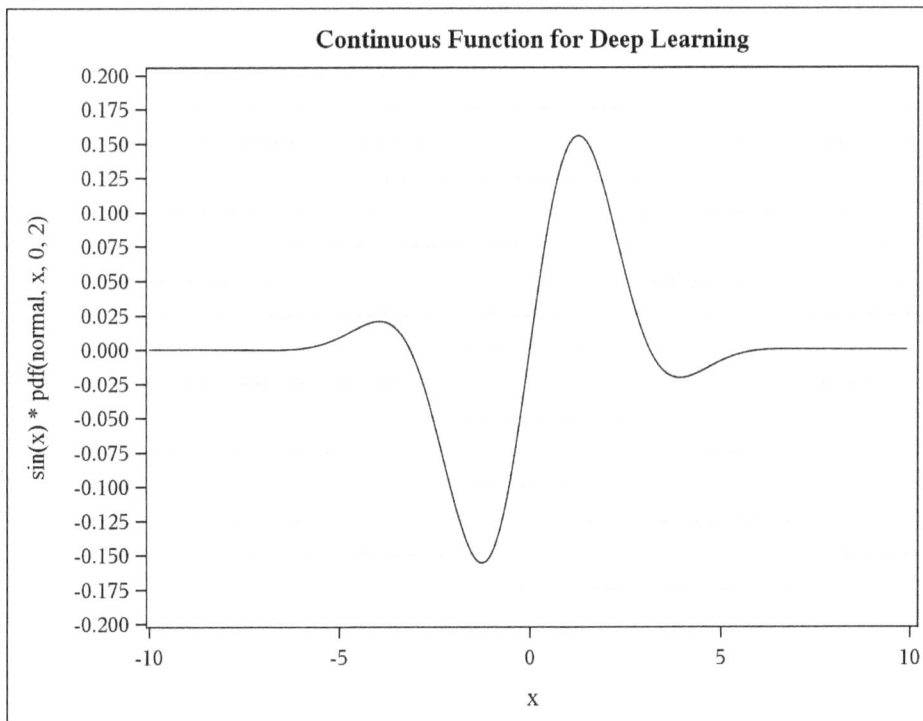

The training (and validation) data is saved in the CASUSER(dl4na) folder (shown here on LINUX):

```
[dl4na@fsnlax05 ~]$ ls -ls ~/casuser/function_training.sashdat
168 -rwxr-xr-x 1 dl4na sas 168384 Nov  2 13:53
/home/dl4na/casuser/function_training.sashdat
```

As we mentioned previously, we now need to separate the training process from the inference process. The training is in dl_function_regression.sas, which is surprisingly simple thanks to our framework.

After creating a CAS session as we did before, we can invoke PROC CAS to build our model:

Program 3.15: Building the Model

```
proc cas;
%include "~/model_io.sas";   /* load_model, etc. */
%include "~/ua.sas";         /* inference, train, etc. */

/* We load the data */
training_table   = "function_training.sashdat";
tables           = setup_training_data({name = training_table}, 80);
training_table   = tables.training;
validation_table = tables.validation;

/* We build the model */
model_name             = "function_model";
number_of_hidden_layers = 8;
define_model(model_name, number_of_hidden_layers);
```

The %include statements enable us to reference our DL framework. We first set up our training and validation data as we discussed before and then create a DL model to approximate any univariate function (the training data tells us the function that we approximate, since we are learning from the data). This code would work for any continuous $y = f(x)$ function.

Hidden within the define_model function is the number of neurons: 1024. You might argue that a smaller (or larger) number would be better. Like many hyper-parameters, this value was selected from experience. A larger network would be harder to train and a smaller network might reduce the accuracy in some cases (or at least make it harder to converge with enough accuracy and no overfitting).

Now that we have a DL model and training data, we are ready to train the model, which is quite easy thanks to the train function:

Program 3.16: Training the Model

```
inputs             = {"x"};
target             = "y";
categorical_inputs = { };
use_gpu            = 1;
max_epoch          = 2000;
train(model_name, training_table, validation_table,
      inputs, target, categorical_inputs, max_epoch, use_gpu);
```

Here again, there is a hidden learning-rate hyper-parameter set to 0.00001 (10-5). It is arguably a small learning rate, but one that works well with Adam and functions with arguments and values close to the origin. As we will see in Chapter 7, we typically normalize our features so that they are close to the origin. In practice, this learning rate works well for us.

After the model is trained, we can save it to disk with one call.

```
save_model(model_name, "CASUSER(dl4na)");
```

Note that `dl_function_regression.sas` contains a few additional lines of code to plot the MSE and how well we're approximating the function of interest (all code that we have seen before).

It is always a good idea to let the training run for a few epochs, because that gives us an early indication of how well or badly the training is doing. So, let's run for 200 epochs, since you already know what 1 epoch would give you:

```
max_epoch          = 200;
```

That should conclude within a minute or two, depending on your hardware.

As you can see in Figure 3.12, the loss function (MSE) for the training set and the validation set steadily go down. Luckily in this training instance, the loss values didn't start at a high value. That is not always the case, and some more epochs might be required to go to 10^{-3} or so.

Figure 3.12: Training Iterations for 200 Epochs

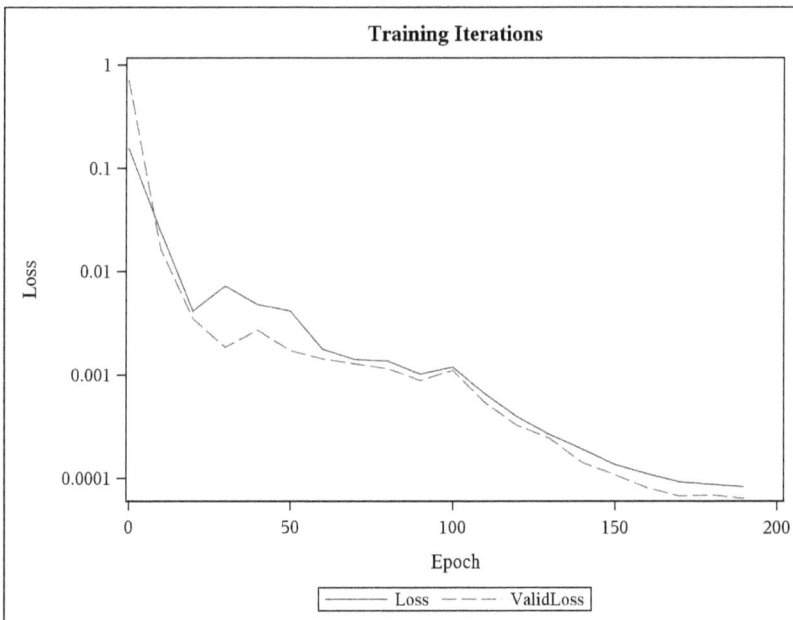

Figure 3.13 shows a plot of the training set true values and predicted values. Although the predicted values are not quite what we would expect, there is some semblance of approximation of the continuous function. We are probably on the right track. Here again, your exact results will probably differ, but they should be similar.

Figure 3.13: Accuracy for 200 Epochs

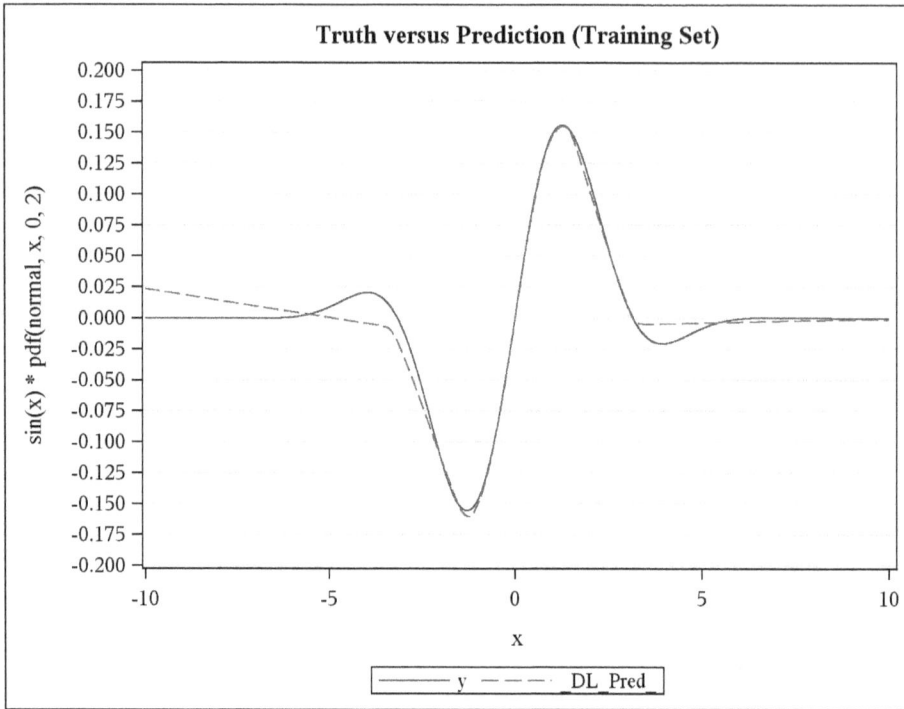

Let's increase the number of epochs so that our optimizer can work harder and build a better approximation of our target function.

We increase the number of epochs ten-fold:

```
max_epoch            = 2000;
```

The training takes about 15 minutes on our hardware. As you can see below, we make good use the use of the GPUs, so if you don't have a GPU, your training time will be significantly longer:

```
+-----------------------------------------------------------------------------+
| NVIDIA-SMI 375.26                 Driver Version: 375.26                     |
|-------------------------------+----------------------+----------------------+
| GPU  Name        Persistence-M| Bus-Id        Disp.A | Volatile Uncorr. ECC |
| Fan  Temp  Perf  Pwr:Usage/Cap|         Memory-Usage | GPU-Util  Compute M. |
|===============================+======================+======================|
|   0  Tesla K80           Off  | 0000:05:00.0     Off |                    0 |
| N/A   58C    P0   128W / 149W |   2248MiB / 11439MiB |     96%      Default |
+-------------------------------+----------------------+----------------------+
|   1  Tesla K80           Off  | 0000:06:00.0     Off |                    0 |
| N/A   38C    P0    78W / 149W |   2248MiB / 11439MiB |     44%      Default |
+-------------------------------+----------------------+----------------------+
```

The plot with the losses for the training set and the validation set can be seen in Figure 3.14. You can also review the accuracy of the training set compared to the truth in Figure 3.15.

We can infer a few interesting points from the plot of the losses:

- The loss is between 10^{-5} and 10^{-6} when we reach 2000 epochs. For many applications, this level of error is satisfactory (for example, a \$1 error on a \$1,000,000 portfolio).
- The losses for the training set and for the validation are steadily going down, so the algorithm is learning at a steady pace. This indicates that our initial guess for the learning rate was good and that Adam is doing a good job converging to a solution that is accurate enough. Also, looking at the curve, we could probably run for more epochs and keep learning.
- The losses for the training set and for the validation set are very close to each other. This tends to indicate that we don't have an overfitting issue and that the network will generalize well (on a data set that the network has never seen before).

Figure 3.14: Training Iterations for 2,000 Epochs

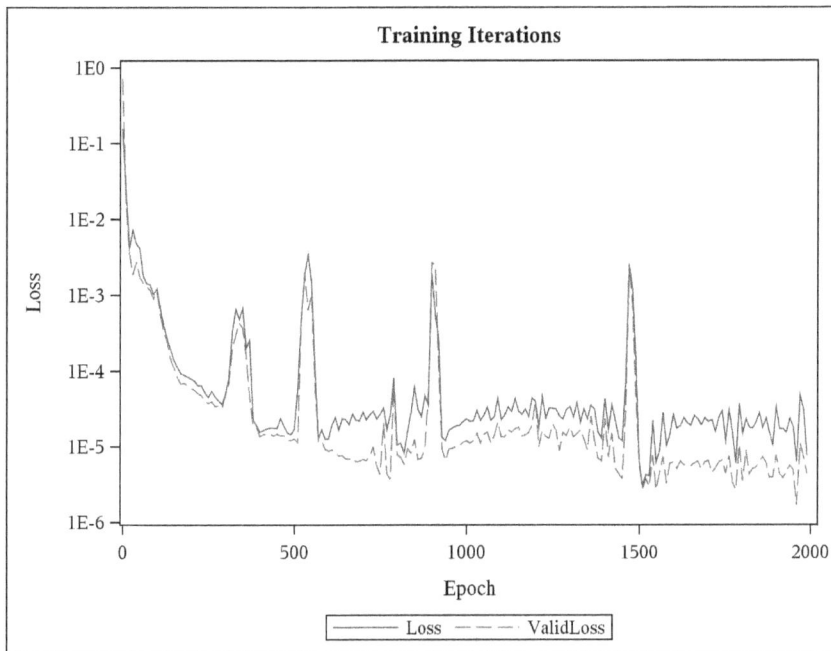

Figure 3.15: Accuracy after 2,000 Epochs of Training

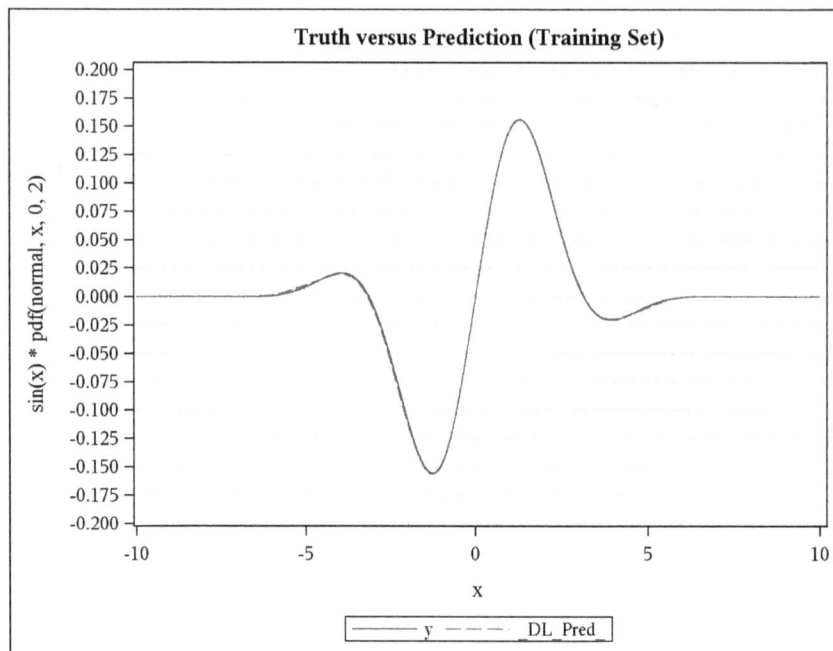

Let's have a look at the scoring with a brand new data set to see if we can replicate the almost perfect match that we see in Figure 3.15.

The scoring code that relies on the saved model can be found in dl_function_score.sas. Thanks to our framework, scoring the model takes only a few lines of code:

Program 3.17: Scoring the Model

```
proc cas;
%include "~/model_io.sas";  /* load_model, etc. */
%include "~/ua.sas";        /* inference, train, etc. */

model_name    = "function_model";
caslib_name   = "CASUSER(dl4na)";
testing_table = "function_testing";
scoring_table = "function_scoring";

/* load the model that was previously trained */
load_model(caslib_name, model_name);

/* Now let's score the test set */
copy_vars = { "x", "y" };
inference(model_name, testing_table, scoring_table, copy_vars);

quit; /* proc cas */
```

As we did previously, we include our DL framework via the %include statements. First we have to load the model that the training saved to disk (in the three tables that we mentioned previously) and then we can run the inference and put the results in function_scoring. The results of the scoring are shown in Figure 3.16.

Figure 3.16: Scoring for 2,000 Epochs

Score Information for FUNCTION_TESTING	
Number of Observations Read	100
Number of Observations Used	100
Mean Squared Error	9.847E-7
Loss Error	4.923E-7

Output CAS Tables			
CAS Library	**Name**	**Number of Rows**	**Number of Columns**
CASUSER(dl4na)	**function_scoring**	100	3

The plot of the accuracy of the scoring is shown in Figure 3.17.

Figure 3.17: Scoring Accuracy after 2,000 Epochs of Training

Score Information for FUNCTION_TESTING	
Number of Observations Read	100
Number of Observations Used	100
Mean Squared Error	9.847E-7
Loss Error	4.923E-7

Output CAS Tables			
CAS Library	**Name**	**Number of Rows**	**Number of Columns**
CASUSER(dl4na)	**function_scoring**	100	3

If you (very carefully) compare the plots in Figure 3.15 and in Figure 3.17, you will notice some subtle differences. For the level of precision that we need in this example, the level of accuracy is satisfactory, but if you need a smaller error, then Cybenko tells you that if your function is continuous, you can find a deep neural network that will accommodate your requirements.

We should point out that the training and validation data are truly different from the test data. The following table shows 20 observations from the training:

Obs	x	y
1	-9.134	-0.00000
2	-4.624	0.01372
3	-4.480	0.01579
4	-4.254	0.01863
5	-3.404	0.01216
6	-2.050	-0.10467
7	-1.256	-0.15572
8	-0.490	-0.09110
9	-0.202	-0.03982
10	-0.050	-0.00997
11	-0.006	-0.00120
12	0.378	0.07231
13	0.706	0.12160
14	6.392	0.00013
15	6.410	0.00015
16	6.652	0.00028
17	8.806	0.00001
18	9.484	-0.00000
19	9.570	-0.00000
20	9.652	-0.00000

In the following table, you can see that the testing data set contains values of x starting at -9.99 and incremented by 0.2:

Obs	x	y
1	-9.19	-0.00000
2	-7.59	-0.00014
3	-7.39	-0.00019
4	-4.79	0.01130
5	-4.39	0.01701
6	-3.39	0.01166
7	-2.79	-0.02596
8	-0.59	-0.10625
9	0.21	0.04135
10	0.41	0.07786
11	2.21	0.08694
12	3.61	-0.01766
13	5.01	-0.00827
14	5.21	-0.00589
15	5.81	-0.00134
16	6.61	0.00027
17	7.61	0.00014
18	8.81	0.00001
19	9.21	0.00000
20	9.41	0.00000

As we mentioned previously, looking at your data (or at least a subset of your data) is always a good idea. This is true before, during, and after training and inference.

Conclusions

In this chapter, we generalized our classification experience with DL networks to regressions. We started with a simple regression using the well-known CARS data set and then moved on to the approximation of a continuous function.

We briefly looked at the on-disk representation of a DL model so that we could separate the training from the inference.

Cybenko's universal approximation theorem gave us a theoretical basis to build a DL framework. We then used that DL framework to easily create a model for a nonlinear function that we could approximate to a high degree of precision.

In the next chapter, we will learn how to organize our analytics as a set of functions that can run in parallel thanks to many-task computing (MTC) methodology. In later chapters, we will build models to approximate analytics that leverage MTC. As promised, we will see that those DL models built on top of MTC reduce the execution times of our models by orders of magnitude.

Chapter 4: Many-Task Computing

In this chapter, we take a slight detour from deep learning (DL) and venture into some supercomputing technologies, namely many-task computing (MTC). We take this detour because we want to take advantage of the latest developments in that field and apply them to the development and deployment of our analytics at scale. By "at scale," we mean in relation to larger problems for more people. The code that we have developed so far was the result of one SAS programmer working alone on the data. We need to expand that single-user experience to a many-user experience and to expand that single-machine experience to a multi-machine experience. To scale up in those directions, we need more tools than the ones that we have used so far.

Most of the material in this chapter is theoretical (we do not actually execute code). We will put the theory into practice in the next chapter when we dive into Monte Carlo simulations.

Before we begin, let's brush up on some terms that are typically not extremely familiar to SAS developers. The concepts introduced in the coming paragraphs are greatly simplified. Our goal is to present enough background in computer hardware to provide a high-level understanding of the choices made by the industry.

A Taxonomy for Parallel Programs

In Chapter 1, we introduced the taxonomy of parallel programs, but with little emphasis on the hardware. Understanding the hardware is essential to understanding why tasks are a much better development paradigm than threads. We will define what a task and a thread are shortly.

As you undoubtedly already know, every computer that runs SAS has a central processing unit (CPU). The CPU is the brain of your computer; it performs logical operations such as tests and arithmetic operations such as addition (von Neumann 1945, Weik 1961). Those operations are performed on the content of memory cells. For decades, CPUs performed one operation at a time (for example, one addition of two numbers). In Flynn's taxonomy (Flynn 1972), those early CPUs were called single instruction, single data (SISD) processors. Often, when software developers refer to a single-threaded program, they are talking about a SISD program or a SISD engine. The `sas.exe` program that you launch when running your SAS programs is a SISD engine. We call it an engine as opposed to a program because it allows for the execution of other SISD programs (your single-threaded SAS programs). The SISD processor is the actual hardware that runs a SISD program (or engine) one instruction at a time. Most modern processors are not SISD processors, as we will discuss shortly.

To coordinate the work between the CPU and devices such as the memory, chip designers rely on an oscillator crystal (also called the clock) that vibrates with some known frequency (the clock frequency). The higher the clock frequency, the faster the computation (more on this later).

Nowadays, CPUs typically contain multiple units that can independently perform logical and arithmetic operations on different data. Those sub-CPUs, if you will, are called cores. A CPU with 4 cores is quite common at the time of this writing, but 64 cores are also possible at the high end. Modern CPUs such as the Intel Xeon are multiple instruction streams, multiple data streams (MIMD) processors. They can run multiple instructions, and each instruction can act upon multiple data elements.

A SISD engine like `sas.exe` can run on a MIMD processor. In fact, a MIMD processor can run multiple SISD engines at the same time: you can start multiple instances of `sas.exe` on your machine. That is possible because the operating system assigns one thread of execution to one `sas.exe` instance, then another thread of execution to another `sas.exe` instance, and so on. Using that methodology, in the next chapter, we will run our SAS code on hundreds of cores at the same time.

So, what is a thread of execution? A thread of execution (or simply a thread) is the smallest unit of execution (sets of instructions) that the operating system can schedule. (A process is typically an accounting unit, not a scheduling unit). MIMD programs that exploit multiple cores typically work with threads, starting them, stopping them, coordinating the work between them. This coordination is usually called synchronization. Synchronization can be defined as the art of preventing more than one thread from writing to the same memory location at the same time. Writing programs that rely on multiple threads of execution is often called multi-threaded programming or programming using multi-threading.

You undoubtedly noticed that we defined synchronization as an art, not a science. We will discuss this choice of words at length in the next section.

There is one last class of processors in Flynn's taxonomy that merits our attention: single instruction stream, multiple data streams (SIMD). A SIMD processor such as a graphics processing unit (GPU) performs the same operation (for example, antialiasing) on multiple data elements (for example, lines). The GPUs are highly optimized to execute SIMD algorithms, and they outperform CPUs by orders of magnitude on SIMD problems. We will see examples of SIMD algorithms later in this chapter.

Now that we have a few concepts defined, let's discuss the evolution of CPUs and GPUs in the last two decades.

We mentioned previously that as a SAS developer, you typically run your program as a SISD application: one statement after another processes one data element after another. In the early days of SAS, from approximately 1980 to 2005, this strategy worked well because the clock frequency of the cores that we used significantly increased from the mid 1990s to the mid 2000s. For SAS developers, this meant that the same analytics would run faster and faster without the developers doing anything (software developers would call this nice state of affairs the "free lunch"). The picture was not always as rosy. When cores run faster and faster, new bottlenecks emerge. For example, once you have solved your CPU problems, your disk access very often becomes a new bottleneck.

But once the data is in memory and the disk can feed the CPUs enough data that they stay busy (not waiting for data), then faster and faster clock rates mean faster and faster performance for your analytics.

Alas for software developers like yours truly, that picture dramatically changed around 2005 or so. Not only did the CPU clock rate plateau, but it started to decrease slightly. Why? Because chip manufacturers found out that to keep the power consumption down and increase the computing capacity, adding more cores made more sense than increasing CPU clock cycles (Ross 2008). For software developers in general and SAS users in particular, this shift in focus from clock rate to integration had a significant consequence: the free lunch was over. In other words, after 2005, SAS users could not get faster analytics performance without seriously altering the structure of their programs. During the next decade, we saw the development of several tools that aimed first at using multiple cores and then multiple machines from SAS programs: SAS/CONNECT, SAS Grid Computing, SAS High-Performance Risk, and CAS, to name only a few.

The goal of these products is to give the SAS developer access to many cores. How many? As of this writing, CPU cores are in the hundreds for one machine, and GPU cores are the tens of thousands for one machine. Before this decade is over, we are likely to be in the range of a thousand CPU cores for one machine.

The products and tools that we mentioned previously enable the SAS developer to write MIMD programs. To support MIMD development, all these products either assume that SAS developers are versed in multi-threading or that the multi-threaded procedures available from SAS already implement the model that they need. (This is not 100% true, as some products, such as SAS High-Performance Risk, enable developers to write some customization of the models by using the FCMP procedure, for example).

It is probably incorrect to generally assume that a SAS developer is versed in multi-threaded development. Simply put, multi-threaded programming presents a significant hurdle that drastically increases in difficulty as the number of cores increases. What might have been manageable with a couple of cores is not necessarily so with hundreds or thousands of cores. So why is it complicated to write multi-threaded programs? In a word, because threads are the wrong level of abstraction (Lee 2006).

Threads are the wrong level of abstraction or the wrong programming paradigm mainly for two reasons that we will briefly discuss in the next section.

Tasks Are the New Threads

The first issue with threads is nondeterminism. If you run a multi-threaded program multiple times, you might get wildly different results. The different results come from one main factor: the scheduling of the threads is up to the operating system, and the state of the overall system is not the same each time you run. As you might imagine, it is very hard for software developers to cope with a seemingly random state in their programs. So the solution is typically to tame the nondeterministic nature of the implementation using synchronization. Critical sections, semaphores, and locks are a few of the tools that help with bringing the level of randomness to a manageable one. The semantics of those synchronization primitives is not trivial. Often one must spend more time on the implementation of the synchronization than on the problem at hand. For SAS developers, this is the wrong thing to do. SAS developers should spend most of their time dealing with their analytical or business issues, not with the constraints that operating system developers impose on them.

The second issue with threads is the work that it takes for the engines or the operating systems to keep the multiple threads synchronized. This overhead typically supersedes the amount of productive work after you reach 50 threads of execution or so. As we stated earlier, 50 threads are not a lot, since we expect cores to be in the thousands fairly soon.

Because multi-threaded programming is so difficult, it is typically left to professional software developers who do this type of work daily. That usually means that SAS users get a multi-threaded procedure such as PROC SORT. However, expecting that a multi-threaded procedure is already available is not a very flexible solution. SAS developers cannot write their own model; they can only parameterize and leverage existing models. Given the complexity of multi-threaded development, this is a compromise that SAS developers have had to live with for years.

What is truly needed is the ability for the SAS developer to write SAS code that uses any DATA steps or available procedures and have the SAS code run on tens, hundreds, or even thousands of threads (cores) without having to embrace the chaos that multi-threading brings to the picture. We will see that task-oriented development provides SAS developers with such a capability. We will also see that task-oriented programming reduces the overhead of synchronization by working at a higher level of abstraction than threads. We will finally see that task-oriented programming helps developers to write better SAS code that is more portable, more maintainable, and much faster.

What Is a Task?

Most programming languages define the concept of a function. Typically, a function is a set of instructions that can take a set of parameters to enable reuse of the instructions with different inputs. For a few years now, people have expected function definitions to define more than one return value, so statements like the following have become commonplace:

```
res1, res2 = function ( in1, in2 )
```

or

```
procedure ( in1, in2, res1, res2 )
```

This level of functional programming has been essential for the software industry to layer software systems on top of software systems in an onion-like fashion to produce very complex systems. Over the years, software developers have realized two important design principles that helped a great deal with multi-threading. First, the input and output argument lists *must* be exhaustive. Second, the input arguments *should* be immutable.

Let's start by considering the input and output arguments lists. When we say that the lists must be exhaustive, we mean that there are no hidden inputs or outputs. Software developers typically call these hidden inputs and outputs "global variables." Writing functions or procedures with hidden inputs or outputs make the development, documentation, and maintenance of those functions a lot harder. Another way to look at this best practice is to realize that a function must not depend on a global state. This principle is true for SISD programs, but it is even truer for MIMD and SIMD programs. If global data structures change every time some function is called, then dealing with the nondeterministic nature of multi-threaded programs becomes a lot harder. So does the synchronization between multiple threads and multiple machines.

In addition to being exhaustive, the input arguments should be immutable. That is to say, one should not change the input argument. Why is that important? Because immutable data structures make it a lot easier to deal with the seemingly random nature of multi-threading and with the overhead of synchronization between multiple threads and multiple machines.

As SAS developers, we are accustomed to the SAS Macro Language, which can heavily rely on macro variables. This reliance on macro variables means that the SAS Macro Language doesn't enforce the two design principles that we just mentioned. To supplement the SAS Macro Language and support those best practices, we introduce the concept of tasks.

A task is a SAS program that defines its inputs and outputs.

This simple definition enables us to run SAS programs on hundreds of cores without even realizing it. We will also see shortly that one benefits greatly by adding one more postulate and stating that a task is a SAS program that does the following:

- defines its inputs and outputs
- treats its inputs as immutable

Following this additional rule allows for the highest level of parallelization.

Let's discuss the rules that transform a SAS program into a task.

Inputs and Outputs

The issue of the granularity of inputs and outputs quickly comes to the surface: Should we use basic types such as integers, Booleans, and doubles? Should we use SAS data sets? Or should we use something else? Since we are dealing with SAS programs, the natural fit for the granularity of the inputs and outputs is a SAS data set. Going down to basic types such as integers is certainly possible, but would potentially greatly increase the burden of synchronization. Like any design decision, using SAS data sets is a compromise. In practice, we have found that compromise to be a desirable one for SAS programs.

SAS data sets for inputs and outputs are not sufficient. Many SAS programs take input files other than SAS data sets (for example, CSV files) and produce output files other than SAS data sets (for example, PDF files). For this reason, we introduce the concept of a data object. A data object is any file that can be used as input or output of a SAS program. As we will discuss shortly, the data object is the unit of synchronization of tasks, so it deserves its own concept (in other words, it is more than just a file).

Before we talk about immutability, we should say a few words about declaring your inputs and outputs. It is a simple rule to enunciate, but not necessarily a simple rule to follow in SAS without the help of some specialized tool. In Chapter 5, we will see how SAS Infrastructure for Risk Management helps a great deal with task-oriented development. At this point, let's simply state that SAS Infrastructure for Risk Management is an MTC platform that finds its origins in the Financial Risk Division of SAS. After a few years in the field, SAS Infrastructure for Risk Management proved itself to be invaluable when it came to writing highly scalable SAS code. SAS Infrastructure for Risk Management is now available to any SAS developer (not just as part of risk products).

Let's briefly discuss some of the implications of the rules and definitions that we just discussed.

Immutable Inputs

In most task development, inputs must not be modified. They are to be considered immutable. This means that if one wants to modify a SAS data set in any way, then one must create a copy of that SAS data set. Modification includes slicing the data set, sorting the data set, truncating the data set, adding or deleting an index file, and so on. Any changes whatsoever to the SAS data set are strictly forbidden. Why? This rule provides a model for efficient synchronization and distribution of the data. This requires a little bit of explanation that we will defer to the next section, when we talk about job flows. In addition, we will see that immutability is an important component of task scheduling in a grid environment.

Based on the introduction of data objects, we need to slightly alter the definition of tasks.

A task is a SAS program that defines its input and output data objects and treats its input data objects as immutable.

We haven't talked about the data model implemented in SAS data sets (the variables of the data set). Once you assign a label to a SAS data set by defining it as a data object, you commit to the data model that is contained in the SAS data set. Once again, the SAS data set is immutable, and that immutability applies to the data model as well. The same concepts apply to a data object in general (for example, the fields of a CSV file). Immutable means immutable—no changes whatsoever.

Now that we understand tasks, we want to combine them to create parallel programs. Creating parallel programs is the topic of the next sections. But before we talk about job flows, let's say a few words about SAS macro variables, since they are a mainstay of SAS development.

Since inputs and outputs are data objects, and since we cannot use hidden inputs or outputs in task development, are we saying that SAS developers cannot use macro variables? Not at all; macro variables are perfectly fine, but they must be passed around as rows of name-value pairs in immutable data objects. For example, consider the following SAS data set:

	CONFIG_NAME	CONFIG_VALUE	CONFIG_VALUE_DESC
1	BASE_DTTM	'31Mar17:00:00:00'dt	...
2	ENTITY_ID	ENTITY_BE	...
3	CONFIG_SET_ID	SAMPLE_34_CONFIGURATION	...
4	GROUP_FLG	N	...

The CONFIG_NAME column contains the name of the macro variable, and the CONFIG_VALUE column contains its value. The CONFIG_VALUE_DESC column is there only for documentation purposes.

What Is a Job Flow?

We will refine this definition later, but for now, let's define a job flow as a set of tasks. It's a set, so there are no duplicates: What would be the point of executing the same task twice? To get the side effects perhaps? But side effects are not allowed, since a task must define all of its inputs and outputs. So no duplicate tasks exist in job flows.

As we hinted previously, a job flow is a parallel program, and the synchronization between multiple tasks is handled via the availability of inputs. Once the inputs of a task are available, the task is ready to execute. If the input of a task is produced as output of another task, then this defines an order for the tasks. In Figure 4.1, you can see a graphical representation of this order. This figure is a partial job flow.

You can see two boxes, one labeled "Filter Zero Coupon Bond Instrument" and the other one labeled "Enrich Interest Rate Curve." Those boxes have circles on top to represent the input data objects and circles on the bottom to represent output data objects. Note that the circle labeled "Filtered zero coupon bond table" is repeated, first as output on the leftmost task and second as input of the rightmost task.

This dependency of outputs and inputs is what allows us to draw an arrow from the first task (on the left) to the second task (on the right). This availability of inputs is the only synchronization that is required when working with tasks.

Figure 4.1: A Partial Job Flow

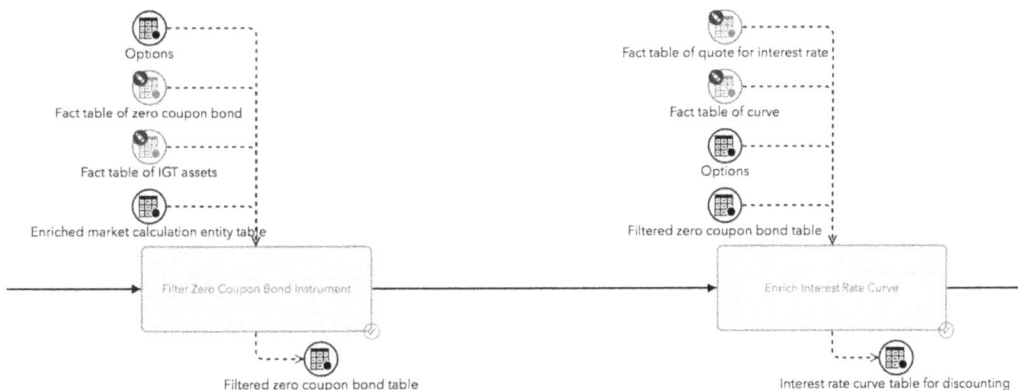

One point that we must emphasize is that during our discussion on tasks, we didn't mention that one task would come after another. This is an important point. During the design and development of the task, you do not need to worry about sequencing (ordering) them. In other words, each task is an island, and the only available bridges are the inputs and outputs. In software engineering terms, a task defines a contract, and to design and implement the task, all you need to focus on is the contract. (This contract for tasks developed in SAS is akin to what is commonly referred to as "interface-based programming"). This isolation or separation of concerns provides support for a federated development process. All SAS developers on the project need only to agree on the task names and their contracts so that they can each develop their tasks independently.

Examples of Job Flows

Now let's have a look at a few job flows to get a better sense of the technology.

The simplest job flow is shown in Figure 4.2. It contains one task with no inputs and one output. That is perfectly OK, but not terribly useful.

Figure 4.2: Hello World

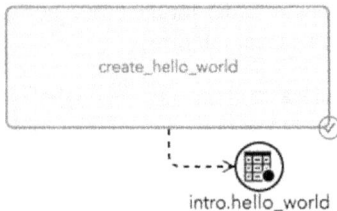

Slightly more useful is the job flow represented in Figure 4.3. The task named "first" reads an input SAS data set and produces an output MK_OPT.a. The task named "second" reads the SAS data set MK_OPT.a as an input and produces the SAS data set MK_OPT.b as an output. The task "second" comes after "first" because of the dependency on the data set MK_OPT.a.

Figure 4.3: A Sequential Job Flow

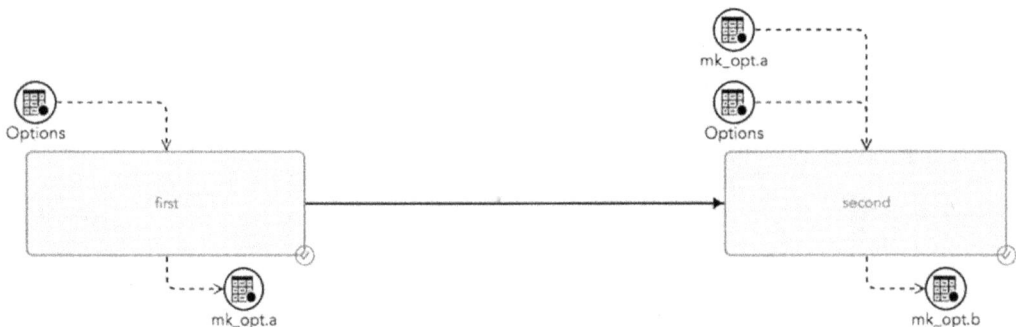

The job flow in Figure 4.4 visually gives you a hint that the two tasks can be executed in parallel. The mere addition of the task into the job flow is what defines the potential parallelism, not the task itself. The task designer or implementer doesn't need to know that the task is being executed at the same time as other tasks. It is not uncommon to have hundreds of tasks in a job flow.

It might look overwhelming at first, but we will see in Chapter 7 how easily one can create complex job flows (including the display of the diagram) using a few SAS statements.

One final note before we get into the next section: the task-oriented programming in which synchronization is done using files and parallelism comes from many tasks is usually called many-task computing or MTC.

Figure 4.4: A Complex Job Flow

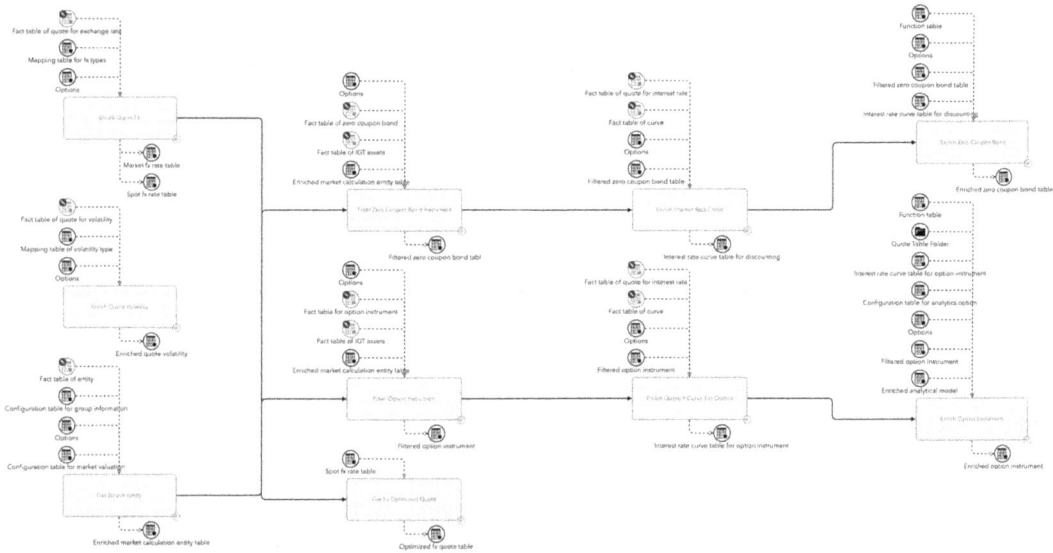

Mutable Inputs

Could we relax the rule of immutable inputs? In a word, yes. Going back to our programming language, we would be talking about a function call like this one:

```
in1, in2 = function ( in1, in2 )
```

Relaxing this rule comes at a cost: the loss of parallelization. In other words, tasks that modify their inputs cannot be executed in parallel. Simply put, this is because one can no longer synchronize on the availability of the inputs; the inputs keep changing. In the following example, we can execute `function1` and `function2` in parallel:

```
b = function1 ( a )
c = function2 ( a )
```

But in the exact same example where the inputs are mutable, `function1` and `function2` must be executed one after the other:

```
a = function1 ( a )
c = function2 ( a )
```

This is true because it is not clear from the code which version of a should be used for `function2`, since there are now two versions of a (before and after the invocation of `function1`). Somehow, the SAS developer must resolve that ambiguity by defining the sequence: `function1` then `function2`, or `function2` then `function1`.

We should point out that in the preceding code, we explicitly declare that table a is mutable by using it on the left side and the right side of the assignment. In other words, even if you want to relax the immutability rule, you still explicitly declare your inputs and outputs. Using that information (table a on both sides), an MTC system such as SAS Infrastructure for Risk Management can appropriately serialize the execution.

The main reason for relaxing the immutability rule is if you have large tables. Either the cost of creating copies defeats the purpose of parallelization of your tasks, or your disk space comes at a premium. Since neither of these conditions is true in this book, we stick to the rule of immutable inputs for all the flows that we use in this book.

Task Revisited

We have now defined a task as a SAS program that specifies its immutable input and output data objects. You might be wondering how this is done. Nothing in the SAS language defines immutable inputs and outputs. Since the definition of the inputs and outputs is part of the contract that we described earlier, we need a way to document and define that contract.

The solution is to use the documentation of the contract to define it. Practically speaking, we use Doxygen to define and document the contract. Doxygen is an open-source documentation tool to generate documentation from source code. (See http://www.doxygen.org for more information.)

Consider the example in Figure 4.5. It contains the source code of a SAS program (fx_conversion.sas) that converts amounts of money from one currency to another.

Figure 4.5: Source Code of the fx_conversion.sas Program

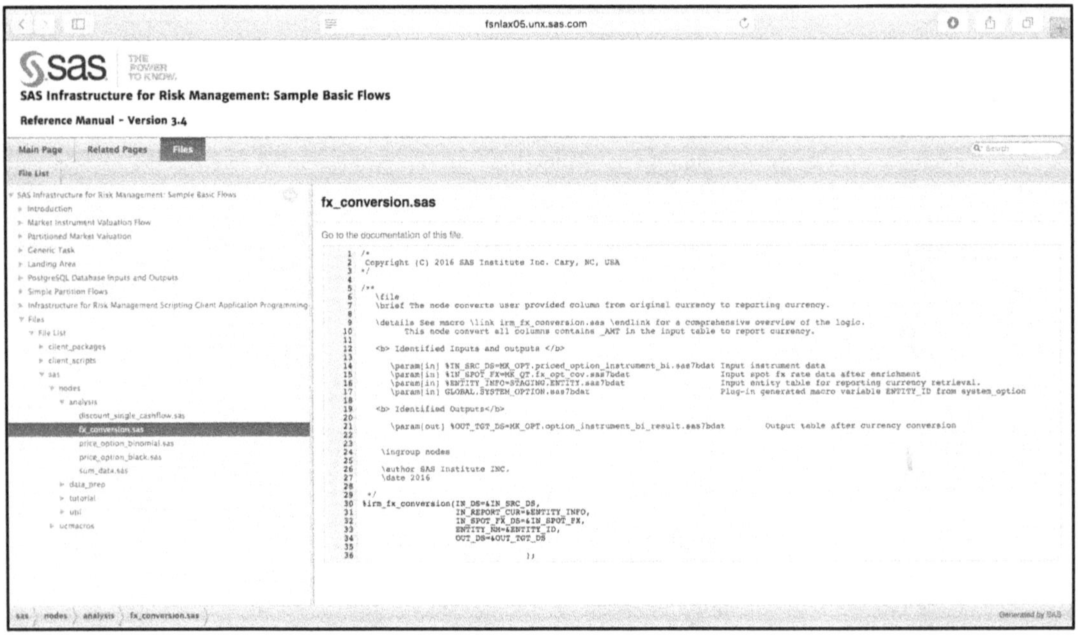

Prior to the actual code (a simple macro call), you can see a comment header with several Doxygen commands. For Doxygen, a command consists of a backslash (\) followed by a keyword. For example, \param[in] MyInput indicates that MyInput is an input parameter. In lieu of the backslash, one can use @, as in @param[in] MyInput.

The comment header from the SAS code in Figure 4.5 enables Doxygen to provide the documentation shown in Figure 4.6. As you can see, not only do we have some documentation of the task, its purpose, its author, and so on, but we also have clear documentation of the contract: the immutable inputs and outputs. In addition to being used by Doxygen to produce the documentation, this interface documentation is also used by an MTC engine (SAS Infrastructure for Risk Management) to order the set of tasks that make up a job flow. As a SAS programmer (and job flow designer), all you need to do is to provide the Doxygen commands. The payoff is that a set of tasks automatically becomes a parallel program that can run on thousands of cores.

Figure 4.6: Documentation for fx_conversion.sas

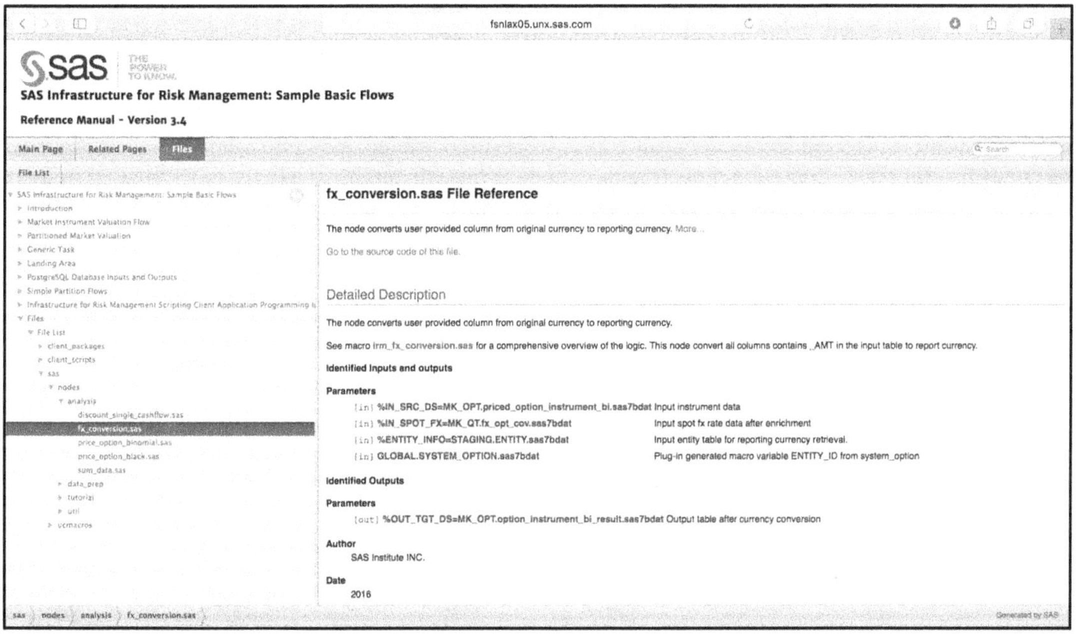

Note that the \param[in] and \param[out] commands contain '%' signs. The '%' sign indicates that the input parameters support substitution. Consider the following example:

```
\param[in] %ENTITY_INFO=STAGING.ENTITY.sas7bdat    Input entity table for reporting currency
```

This code indicates that one could call the task with the default arguments, as follows:

```
%fx_conversion(
          t_in_src_ds    = mk_bond.filtered_zero_coupon_bond.sas7bdat,
          t_in_spot_fx   = mk_qt.fx_opt_cov.sas7bdat,
          t_out_tgt_ds   = mk_bond.enriched_zero_coupon_bond.sas7bdat);
```

Or one could substitute an alternate ENTITY table (`staging.entity2.sas7bdat`):

```
%fx_conversion(
        t_in_src_ds    = mk_bond.filtered_zero_coupon_bond.sas7bdat,
        t_in_spot_fx   = mk_qt.fx_opt_cov.sas7bdat,
        t_entity_info  = staging.entity2.sas7bdat,
        t_out_tgt_ds   = mk_bond.enriched_zero_coupon_bond.sas7bdat);
```

Before we discuss how to package tasks so that they can be distributed and promoted to multiple environments, we need to briefly discuss the encapsulation of SAS code that tasks provide.

As we've discussed earlier, a task defines its immutable inputs and outputs. Another way to look at the task is as a labeled SAS program that, given a set of inputs, produces a set of outputs. There is absolutely no mention of how the task achieves such a thing. In its implementation, the task could define and exploit many data structures such as temporary tables (for example, SAS WORK), local and global macro variables (passed around as SAS data sets), hash tables, and so on. But nowhere in the definition of the task is that internal implementation required. That is encapsulation, one of the three characteristics of an object-oriented system. The other two are inheritance and polymorphism. We will see shortly that tasks exhibit at least some aspects of inheritance and polymorphism.

One more thing to note before we talk more about parallelism is that a set of inputs that produces a set of outputs is what feeds deep learning. Furthermore, when we combine tasks into job flows, we still have a set of inputs that produces a set of outputs (the intermediate outputs might be left out of the picture in this case). In other words, job flows are a nice fit for enabling a deep neural network to learn. We will implement a Monte Carlo simulation with these facts in mind in Chapter 7.

Partitioning

We have seen that by breaking down a problem into tasks, we can gain in performance by using more cores, since some tasks can execute in parallel. This tactic is called *task partitioning*: you slice your problem into tasks that can run in parallel as soon as their respective inputs are available. Task partitioning is an example of a MIMD algorithm in which all tasks execute independently on their unrelated data.

Some problems can be divided into sub-problems at the data level with little to no effort. Those problems are sometimes called embarrassingly parallel problems or perfectly parallel problems. For example, if you need to price one million securities, you can price them independently of each other. Scoring is another example of an embarrassingly parallel problem. These types of problems are in fact ideally suitable for SIMD processors and SIMD engines. In MTC, you simply slice up the inputs of a task into multiple data sets (all have the same data model), fire up as many tasks as you have slices, and then recombine the outputs into one single output. When multiple tasks work on data partitioned in the same way (for example, by the same variable), we don't recombine the outputs to avoid unnecessary processing. Recombining the outputs is required only when at least one task that uses that output as input doesn't use data partitions.

This division of labor based solely on the data is called *data partitioning*. Note that nothing prevents you from partitioning by task and then partitioning by data in the same job flow to gain as much parallelism as possible. We will look at practical examples of data partitioning in the next chapter.

Some problems are not perfectly or embarrassingly parallel but can benefit from partitioning. Sorting is a good example. Sorting can greatly benefit from the performance of a SIMD processor (Satish et al. 2008). One way to deal with this type of problem is to take advantage of the encapsulation that tasks provide. You embed the SIMD implementation inside the task (so that no data partitioning is visible on the inputs or outputs of the task) by writing a CUDA implementation of the task (Bequet and Chen 2017).

In the next section, we discuss a possible methodology for packaging tasks so that they can be easily distributed and modified.

Federated Areas

As we discussed earlier, the clear contract that tasks provide is conducive to a distributed (or federated) development model. For example, developers in Europe can code per the contract defined by developers in the USA. However, simply having the specification of the behavior is not enough. We need a well-organized structure for the code, its inputs, and its outputs. Federated areas (FA) are a tool to manage job flows, tasks, data objects, and their lifecycles.

Specifically, an FA is a folder structure like the one shown in Figure 4.7.

Figure 4.7: A Federated Area

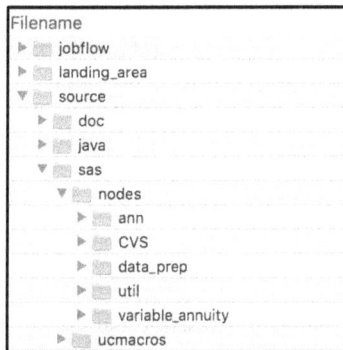

```
Filename
▶ 📁 jobflow
▶ 📁 landing_area
▼ 📁 source
    ▶ 📁 doc
    ▶ 📁 java
    ▼ 📁 sas
        ▼ 📁 nodes
            ▶ 📁 ann
            ▶ 📁 CVS
            ▶ 📁 data_prep
            ▶ 📁 util
            ▶ 📁 variable_annuity
        ▶ 📁 ucmacros
```

Let's first focus our attention on the top-level folders:

- **jobflow**

  In this folder, we put the definitions of the job flows that we encountered earlier. We will see in the next chapter how to create job flow definitions using SAS code.

- **landing_area**

  This folder contains the inputs of the tasks that are not outputs of other tasks. In other words, the landing area contains the data objects that are only in the inputs of tasks. Typically, this folder is populated by an ETL process.

- **source**

  This folder contains the code that implements the tasks. This is usually SAS code, but can potentially be any other language, such as Java, C, or CUDA (CUDA, which we will revisit in Chapter 6, is the development language for NVIDIA GPUs). In this book, we mostly focus on SAS code.

Under the `source` folder, we find a `sas` folder and a `doc` folder. The `sas` folder contains the SAS code for your tasks arranged in two subfolders:

- `nodes`

  The folder is called `nodes` but could as well have been called `tasks` (the folder name refers to the nodes of the directed graph that is used to schedule tasks to as many cores as possible). You will notice that the `nodes` folder contains subfolders to organize your SAS code (tasks) any way you see fit (in this example, we see the folders for a machine learning application). This organization is somewhat analogous to packages that you encounter in other languages such as Java.

- `ucmacros`

  The `ucmacros` folder enables you to package all the macros called by the tasks of your federated areas (more on this later).

Federated areas are more than a way to package your task implementations and their input data. Things become interesting when you have more than one federated area, because their content can participate in an inheritance relationship.

Consider the following case. In the USA, insurance regulations are defined on a state-by-state basis (each state has its own insurance commissioner). The rules and regulations that apply to Alabama and California are somewhat different, but there is some commonality between those states. One can purchase life and auto insurance in both states, but the minimum liability for auto insurance could be different (more complex differences exist). Developers at a large insurance company that relies on SAS for its analytics could design federated areas as follows:

- one common federated area for the entire country (FA.USA)
- several federated areas for different regions (FA.EAST, FA.WEST)
- another 50 federated areas for the states (FA.ALABAMA, FA.CALIFORNIA)

One might suppose that the FA.ALABAMA would contain everything that is in FA.EAST and FA.USA, but that is not a very palatable solution for several reasons. First, duplication is rarely a good thing. With one version of the truth things are easier and cheaper to design and maintain. Second, the developers working on the different federated areas could no longer work independently by just following the contract embedded in the tasks. They would have to coordinate their efforts and releases. A much better solution is for federated areas to support inheritance in a tree structure like the one in Figure 4.8.

Figure 4.8: Inheritance of Federated Areas

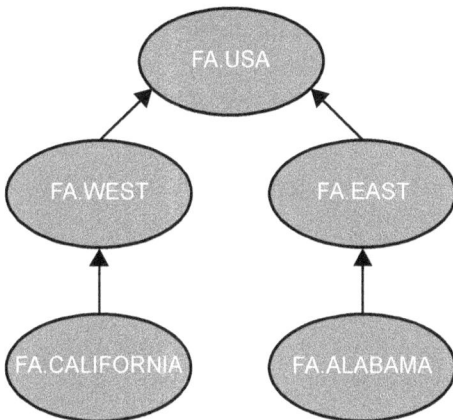

This diagram illustrates the following inheritance characteristics of the federated areas:

- FA.EAST and FA.WEST inherit from FA.USA
  In other words, when analytics are run, the artifacts (tasks, job flows, and data objects) of FA.EAST are merged with the artifacts of FA.USA. When an artifact is defined in FA.USA and FA.EAST (for example, a data object containing the minimum liability), the one in FA.EAST takes precedence.
- FA.ALABAMA inherits from FA.EAST
  This means that it consequently inherits from FA.USA as sell.
- FA.CALIFORNIA inherits from FA.WEST

If you're familiar with object-oriented development, you have probably noticed that this inheritance of federated area artifacts is like inheritance of class (and interface). Note that in the case of tasks, there is only one unnamed method: the SAS code contained in the `<task-name>.sas` file. It is a single inheritance: one FA can have only one parent. This concept of inheritance is very powerful for federated development. It not only avoids duplications, but also enables you, during development, to keep the federated areas that are present in deployments and then create a federated area with just the code or the data that you want to modify. There is no need to copy the artifacts that you're not working on.

We have seen that the landing area contains the inputs of the tasks, but what about the outputs? Where do they go? They go to a folder called the persistent area.

Persistent Area

Until now, we have used the words job flow to mean both a job flow definition and a job flow instance:

Job flow definition
  The list of tasks and their inputs and outputs that constitute the job flow (also known as a parallel program). Since the task is inseparable from the contract, it is included in the job flow definition.

Job flow instance
  An actual run of the tasks that includes the outputs of all the tasks that have run. The log file is considered an implicit output (all tasks have one, which is possibly empty). One can have many

instances of a job flow definition. They typically differ by their inputs in the landing area. Here again, you can see the analogy with object-oriented systems, where the class is the definition and the objects are the instances.

When running the code, this distinction becomes very important, and we need to give an explicit identifier to the job flow instance: the instance ID, which is simply a long integer.

Figure 4.9 shows an example of a persistent area. The `1305712421` folder name is the instance ID. Other folders, such as `datalib`, are the librefs of the outputs of tasks. Note that there are several `mvatask*.log` files. They contain the SAS logs of the execution.

Figure 4.9: A Persistent Area

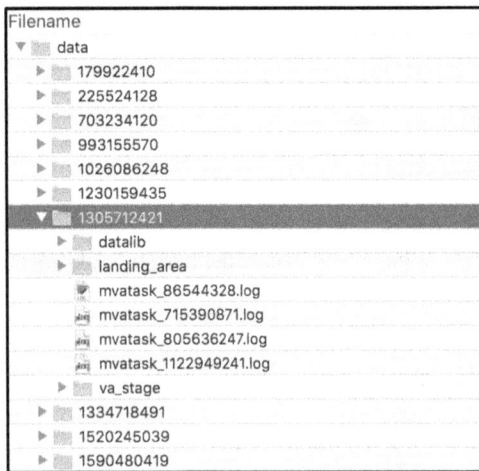

```
Filename
▼ ▦ data
   ▶ ▦ 179922410
   ▶ ▦ 225524128
   ▶ ▦ 703234120
   ▶ ▦ 993155570
   ▶ ▦ 1026086248
   ▶ ▦ 1230159435
   ▼ ▦ 1305712421
      ▶ ▦ datalib
      ▶ ▦ landing_area
        ▦ mvatask_86544328.log
        ▦ mvatask_715390871.log
        ▦ mvatask_805636247.log
        ▦ mvatask_1122949241.log
      ▶ ▦ va_stage
   ▶ ▦ 1334718491
   ▶ ▦ 1520245039
   ▶ ▦ 1590480419
```

One folder should pique your interest: `landing_area`. Isn't that where the inputs are? Why do we need a `landing_area` folder in the persistent area? The reason is simple. Once you have created an instance of a job flow, you should be able to run it with different inputs to try some variations on your model. What if the conversion rate were higher? What if it were lower? What if you had more cash on hand? Those "what-ifs" typically come in the form of alternate versions of one or more of your input tables, so where would you put those additional versions? Since those additional versions are tied to an instance of the job flow, putting them in the persistent area seems like a natural choice.

Note that creating an additional version of a SAS data set and rerunning the job flow with that additional version as input is not the same as modifying the input inside the task. The rule about immutability applies to the task. In the implementation of the task, you must treat the inputs as immutable so that when the task executes, the inputs are left unchanged. What happens outside the definition of the task is an entirely different matter. Specifically, in SAS Infrastructure for Risk Management, there are three ways to modify the inputs when a task is not running:

1. When the server is not running, load or copy data into the landing area.
   When the server comes back up, SAS Infrastructure for Risk Management automatically identifies which job flows are out-of-date and consequently need to be re-executed.
2. When the server is running, load or copy data into a special folder called the input area.
   We will see an example of this functionality called live ETL (Extract, Transform, and Load) in

Chapter 5. Here again, SAS Infrastructure for Risk Management automatically identifies which job flows are out-of-date and consequently need to be re-executed.

3. On a job flow instance by job flow instance basis, use the GUI or the REST API to upload a new version of the data.

Finally, note that a typical MTC platform (such as SAS Infrastructure for Risk Management) uses a reference counter to determine when to discard the versions that are no longer needed. We will discuss this more later when we talk about object pooling.

Before having a closer look at performance, let's first discuss some caveats and pitfalls to avoid when working with MTC.

Caveats and Pitfalls

In this section, we briefly discuss both best and bad practices of MTC, and what will go wrong if we don't follow the rules that we discussed earlier. Before we discuss specifics of what should or shouldn't be done, note that any MTC platform that runs on top of SAS has to live with the power and flexibility of SAS. Basically, you can do nearly anything in your SAS code, so the burden of following the rules falls largely on the SAS programmer. Caveat emptor!

Another general word of caution is that the MTC rules that we discussed are designed to avoid the chaos of multi-processing and multi-threading. Not following the rules guarantees the return of the chaos, but typically in a sneaky way: the program runs fine on a machine that happens to have 4 cores, but fails on a machine or a grid with 256 cores.

Not Declaring Your Inputs

Not declaring your inputs is a very bad idea. Several key functionalities of an MTC platform (or engine) such as SAS Infrastructure for Risk Management will be severely degraded in a chaotic manner:

1. The execution of your task will happen at the wrong time.
   Since the only way for MTC to know when your task can execute is to look for the availability of the inputs, that decision would be made on the wrong information.
2. The execution of your task will fail.
   This failure typically happens because the input of your task is still being generated by another task. The effect in SAS code is usually a "data set is locked" error. The same issue can occur in a grid environment because the file is still being copied.
3. The execution will produce the wrong data.
   This issue is the same as the previous one, except that you happen to be "lucky" (unlucky, really) and avoid the locking issue. This means that the task will compute your analytics with the results generated by a different version of the inputs. This is probably the cardinal sin for a data scientist.
4. Job flow instances will not be correctly marked out-of-date.
   When you load new data, SAS Infrastructure for Risk Management will tell you which job flow instances are no longer up-to-date. Without correctly identifying the inputs, this functionality can no longer be provided.
5. Temporary libraries won't work correctly.
   This feature is described later in the "Data-Object Pooling" section.
6. Many other issues can occur because of faulty inputs leading to faulty outputs.

Now let's focus on what happens if you break the immutability rule.

Not Treating Your Immutable Inputs as Immutable

In our discussion about immutability earlier, we stated that you would indicate that a table is mutable by including it in the inputs and in the outputs. For example, the following declaration is fine because you explicitly state that table a is mutable:

```
\param[in]    STAGING.a.sas7bdat    First input
\param[in]    STAGING.b.sas7bdat    Second input
\param[out]   STAGING.a.sas7bdat    First output (also an input)
\param[out]   STAGING.c.sas7bdat    First output (not an input)
```

What would be problematic is the following declaration for code that modifies table a:

```
\param[in]    STAGING.a.sas7bdat    First input
\param[in]    STAGING.b.sas7bdat    Second input
\param[out]   STAGING.c.sas7bdat    First output (not an input)
```

That is an extremely bad idea. Several key functionalities of an MTC platform (or engine) such as SAS Infrastructure for Risk Management will be severely degraded in an even more chaotic manner than in the previous section:

1. The execution of some tasks will fail.
    This typically happens because the input of your task is being used by another task while the offending task is still running. The effect in SAS code is usually a "data set is locked" error. The same issue can occur in a grid environment because the file is still being copied.

2. The execution will produce the wrong data.
    This issue is the same as the previous one, except that you happen to be "lucky" and avoid the locking issue. This means that the task will compute your analytics with the results generated by a different version of the inputs.

3. Job flow instances will not be correctly marked out-of-date.
    In this case, the MTC platform cannot determine the correct state of the inputs.

4. Many other issues can occur because of faulty inputs leading to faulty outputs.

Not Declaring Your Outputs

The main consequence of not declaring the outputs of a task is that those outputs cannot be used as inputs by other tasks. Nothing bad will happen except in the following cases:

*   Not declaring your outputs brings you back to a state where your inputs are not declared so that you can use them. In that case, the problems that we discussed previously will surface.

*   Your undeclared outputs are declared as outputs of another task. In that case, a "data set is locked" error is likely (or even worse, the two tasks might produce different results).

By now, you should be convinced that correctly declaring the immutable inputs and the outputs of every single task is essential. Finding the exact list of your inputs and outputs might not always be trivial, especially if you're refactoring some code. However, there is help available with the Source Code Analysis procedure (PROC SCAPROC), which lists all your inputs and outputs when your code executes. A detailed discussion of PROC SCAPROC can be found in Chapter 60 of the *Base SAS 9.4 Procedures Guide*, available at http://documentation.sas.com/api/docsets/proc/9.4/content/proc.pdf.

Finally, let's mention this important practical rule when writing SAS code for tasks: if you're declaring a libref, you're probably doing something wrong. This is because declaring a libref implies that your MTC platform couldn't generate the LIBNAME statement from the declaration of your inputs and outputs, and consequently you didn't declare them correctly.

Let's now focus our attention on the MTC performance of a multiple-machine deployment.

Performance of Grid Scheduling

RAM is much faster than disk input/output (I/O). To take advantage of the speed of RAM during disk I/O, modern operating systems such as LINUX or Windows rely on a disk cache or page cache (Love 2010).

When all tasks in a job flow run on one machine, the output of a task is typically in RAM by the time that the next task is ready to use it. Essentially, there is little to do to gain the performance advantage of the disk cache in a job flow. You just give a higher execution priority to tasks that use the output of tasks that just finished and simply let the OS do its work. This is not true when the tasks of a job flow don't run on a single machine. If the first task of our example in Figure 4.3 runs on one machine and the second task runs on another machine, the MK_OPT.a data set will not be in memory when the second task is ready to run. Even worse, the output data set of the first task must be transferred to the machine of the second task over the network. All this data movement significantly slows down the execution of tasks to the point that adding more machines typically slows down the execution of the job flows. To avoid this frustrating situation, we can select the machine to run tasks to minimize the data movements and to optimize the usage of the disk cache. Presenting the details of the scheduling is outside the scope of this book, but it is worth noting here that without the knowledge of the inputs and outputs, we would not be able to minimize the data movements and optimize the usage of the disk cache (Zhang et al. 2016).

We can see that relying on MTC provides an easy solution to a complicated problem (Tanaka and Tatebe 2014).

Data-Object Pooling

We have spent the better part of this chapter describing task-oriented programming where the work is organized only in tasks that define their immutable inputs and outputs. Another way to look at these statements is to say that all outputs of a task in task-oriented programming are computed by a succession of inputs, tasks, and outputs (of other tasks).

More precisely, the succession of tasks form a directed acyclic graph (DAG) where the predecessors and successors are defined by the availability of data objects. A task comes before another when it produces outputs that are used as inputs by the other. This means that for each output in the system, we can remember the subset of the DAG that was used for the computation. That information is a simple ordered set of data objects and task identifiers. So whenever we need to schedule a task for execution, we have an easy (and fast) way to find out whether that output already exists after a lookup for the ordered set. This is the essence of data-object pooling: the ability to quickly determine whether a task needs to be computed or not.

The savings that are provided by data-object pooling are twofold: speed and space.

Obviously, not executing the task at all can provide tremendous performance improvements if the lookup is much faster than executing the task. If the lookup is in fact an in-memory hash table lookup, then it will beat the combination of disk access and computation every time.

The space savings are also significant, because if we don't need to run the task to create the output, we can simply create a file system link to the existing output and be done. The savings come from the fact that storing the link is much more efficient than storing the information.

In practice, the outputs are managed using a reference counter and can be discarded when the reference counter reaches 0. Obviously, this can hardly be done by hand, so one needs a software system to manage data-object pooling. We should mention that not correctly declaring your immutable inputs and outputs will lead to catastrophic failures. Data-object pooling will make the decision based on erroneous graph paths and the overall state of the system will be unpredictable.

Data-object pooling also provides an ancillary savings that comes from the fact that by reusing outputs that were previously calculated, the disk cache is better used. Fewer page faults lead to more I/O operations performed in RAM. Note that a page fault occurs when the software accesses data that is not in memory and needs to be loaded from disk.

How significant are the savings? With the job flows defined in SAS Firmwide Risk for Solvency II, we have measured both a performance and a disk space improvement close to one order of magnitude with typical user workflows. The more users there are and the more often the flows are executed or modified, the better the savings. We should also point out that if your analytics are deployed in a third-party data center, you will probably pay for renting the disk space just as you pay for the cores. The big difference between the cores and the disk space is that once the job flow is done executing, you can recycle the cores, but the disk space remains in use for a longer time (and you keep paying for it). In this case, less disk space can amount to big savings. In addition, if you don't need the intermediate results, then a DAG-oriented product such as SAS Infrastructure for Risk Management enables you to put some outputs in a temporary libref. The content gets deleted as soon as the data set is no longer needed. This is only possible because of MTC. Without knowing all the inputs of all the computations, one cannot readily tell if some are no longer needed.

Finally, there are no security risks introduced by data-object pooling, because users have no idea that their outputs were in reality computed by someone else. The only thing that the end user notices with data-object pooling is that the system becomes faster and faster as it is used. (If one pays close attention, the date and time of the SAS log will reflect the true time of execution of the task). Another way to state this fact is to say that the machine automatically learns from the data consumed and produced by the user.

Here again, relying on MTC provides an easy solution to a complicated problem.

Portable Learning

In the previous section, we discussed the advantage of data-object pooling in helping to avoid executing some tasks, because the same task with the same inputs had been executed before and thus the outputs were readily available. The more expensive the task, the more benefit we derive from data-object pooling. What is true for one task is true for many, so the computation of an entire job flow might be avoided if the results are readily available. Here again, the bigger the job flow, the bigger the savings in execution time and disk space.

One workflow that can greatly benefit from data-object pooling is deep learning. In DL, we first train the network, and then we exploit it for inference or scoring. The training part might take a long while. With data-object pooling, the system can automatically identify, in a job flow like the one shown in Figure 4.10, whether the training part needs to be executed or not. This determination can be done without any manual intervention from the SAS developers. Thanks to data-object pooling, this optimization is taken care of automatically. In addition, in the example of Figure 4.10, we are talking about one system where the job flow is installed and running. However, if you need to promote your learning to another machine, this works as well. Note that you can do the training with the training subflow on one system, and then do the inference on another system. The job flows don't need to be the same on both systems. Only the tasks and their inputs and outputs need to be common to the training and inference job flows, since data-object pooling works at the task level, not at the flow level.

Figure 4.10: A Deep Learning Job Flow

Conclusion

In this chapter, after a brief introduction to computer hardware, we have introduced many-task computing: a development methodology that relies on breaking down problems into small SAS programs that define a contract in terms of the outputs that they produce based on the inputs. We have seen that the inputs and outputs were not necessarily SAS data sets, so we called them data objects. We have discussed how data objects could be partitioned in some cases to gain even greater parallelism.

Armed with data objects, we could define an automatic synchronization model so that many of our tasks could run on as many cores as we have at our disposal. We have contrasted this automatic synchronization with the more traditional multi-threading model that is ill-suited for SAS developers and brings chaos into their work.

We then saw how to write SAS code following that model: document the contract and then abide by the contract. We also emphasized that correctly defining the contract and complying with it was essential for correctness and smooth operations.

After that, we turned our attention to packaging and distributing analytics written using the MTC paradigm.

We concluded the chapter by looking at complicated problems such as grid scheduling. We discussed how MTC provided the necessary information for multi-machine schedulers to effectively use the disk cache for optimal performance. We also briefly discussed how MTC provides a nice framework for DL at large.

Now that we have the MTC concepts under our belt, we can start using them to design and deploy analytics at scale. That is the main topic of the remainder of this book, starting with the next chapter: "Monte Carlo Simulations."

Chapter 5: Monte Carlo Simulations

In this chapter, we briefly introduce Monte Carlo simulations. Monte Carlo simulations are an important tool in many disciplines, including finance, manufacturing, and physics. Broadly defined, Monte Carlo simulations are a class of algorithms that rely on random samplings to estimate stochastic and deterministic values. Monte Carlo simulations are especially valuable when a closed-form solution is not available.

By the end of this chapter, you will understand Monte Carlo simulations, random walks, and how to run simulations at scale with SAS.

Monte Carlo or Las Vegas?

In 1733, the French naturalist and amateur mathematician Georges-Louis Leclerc, Comte de Buffon proposed the following problem during a lecture at the Académie Royale des Sciences (Buffon 1733):

> *Let's say that you have a hardwood floor (parquet) of equidistant boards. If you randomly throw needles on the floor, what is the probability that a needle will land between two boards?*

This situation is summarized and idealized in Figure 5.1.

Figure 5.1: Buffon's Needles

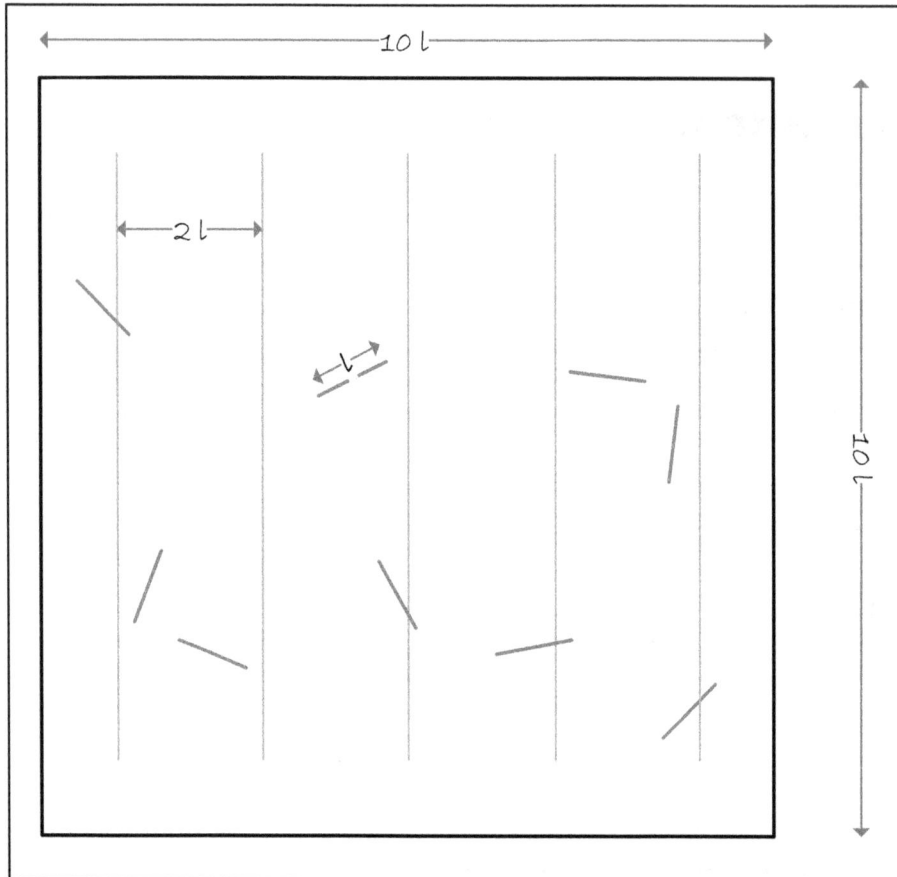

When Buffon posed the problem in 1733, it had nothing to do with π, at least at first glance. A few years later, in 1777, Buffon solved the probability problem (Buffon 1777), but, alas, his solution contained an error. That error was later addressed by Pierre-Simon de Laplace in 1812 (Arnow 1994), and the probability was stated as follows in the case of $d > l$, where d is the width of a board and l is the length of a needle:

$$P = \frac{2\,l}{\pi\,d}$$

Note that in Figure 5.1, $d = 2\,l$. Laplace is believed to be the first one to propose to estimate π with Buffon's experiment, thereby giving birth to Monte Carlo simulations, which are defined as follows:

A methodology to solve deterministic or probabilistic problems using a simulation of random variables.

In the case of π, the problem is deterministic. We will see an example of a stochastic problem in the next section, "Random Walk."

The most famous manual attempt at performing the simulation suggested by Laplace was reportedly conducted in 1901 by Lazzarini (1901) who claimed to have approximated π with six decimals. That work was later debunked by several mathematicians, including Lee Badger (1994). Note that in the context of our discussion, the point of Buffon's needles is not to be the best estimator of π, but to serve as a good introductory example of a Monte Carlo simulation. For example, Ramanujan's formula is a much better estimator of π:

$$\frac{1}{\pi} = \frac{2\sqrt{2}}{9801} \sum_{k=0}^{\infty} \frac{(4k)!\,(1103 + 26390k)}{(k!)^4\,396^{4k}}$$

It was not until the Manhattan Project that we saw the modern version of Monte Carlo simulations.

Stanislaw "Stan" Ulam was a Polish mathematician who worked on the Manhattan Project with people such as Enrico Fermi and John Von Neumann. Stan was recovering from an illness and trying to kill time by playing solitaire. Apparently, he wasn't very good at it, and as a mathematician, he wanted to understand why. Since he couldn't find a closed-form solution, he had the idea of running simulations. He didn't have a computer, but his friends John Von Neumann and Nicholas Metropolis had access to the Electronic Numerical Integrator and Computer (ENIAC). Von Neumann and Metropolis recognized a good idea when they saw one and used the methodology to simulate neutron transport, the motions and interactions of neutrons with material. That experiment on the ENIAC is reportedly the first Monte Carlo simulation on a digital computer (Metropolis and Ulam 1949).

Nowadays, we have easy access to the modern version of the ENIAC, so let's conduct a modest Monte Carlo simulation of Laplace's suggestion to use Buffon's needles to estimate the value of π.

Here is a straightforward, almost naïve implementation in SAS, using the FCMP procedure (PROC FCMP):

Program 5.1: A Modest Monte Carlo Simulation

```
/*
   The needle is fully defined by its length (1 unit),
   by the (x, y) coordinates of its center, and by theta,
   the smallest angle between the needle and the horizontal.
*/
subroutine get_needle(
   x,       /* OUT The x coordinate of the center */
   y,       /* OUT The y coordinate of the center */
   theta); /* OUT The rotation of the needle     */
   outargs x, y, theta;
   x     = rand("UNIFORM") * 10;
   y     = rand("UNIFORM") * 10;
   theta = rand("UNIFORM") * constant("pi") * 0.5;
endsub; /* get_needle */
```

Once we have a needle, we generate a lot of them and compute how many intersect our vertical lines (you'll notice that there is no need to compute the Y coordinates):

Program 5.2: Generating Needles

```
%let nb_needles = %sysget(nb_needles);

data _null_;
  intersect = 0;
  do ndx = 0 to &nb_needles.;
    x = 0;
    y = 0;
    theta = 0;
    call get_needle(x, y, theta);
    xmin = x - 0.5 * cos(theta);
    xmax = x + 0.5 * cos(theta);
    do lndx = 1 to 9 by 2;
        if (xmin < lndx) and (xmax > lndx) then
        do;
           intersect = intersect + 1;
        end;
    end;
  end;
  estimate = &nb_needles. / intersect;
  put intersect= ' PI estimate is' estimate=;
run;
```

Note that it is possible to write much faster versions of this algorithm, as Rick Wicklin describes on his DO Loop blog (https://blogs.sas.com/content/iml/2012/01/04/simulation-of-buffons-needle-in-sas-2.html).

If we run our estimator with a few draws, then we get a poor estimate of π, but not an abysmally bad one:

```
$ sas buffon.sas -set nb_needles 1000 -stdio
intersect=318  PI estimate isestimate=3.1446540881
```

If we increase the number of draws, the estimate improves a bit:

```
$ sas buffon.sas -stdio -set nb_needles 1000000
intersect=3183043644  PI estimate isestimate=3.141650749
```

You'll notice that this is nowhere close to what Lazzarini claimed to have achieved.

One final word before we consider random walks: the name Monte Carlo is rumored to come from Stan Ulam's uncle, who continuously borrowed money from relatives because he "just had to go to Monte Carlo" (Metropolis and Ulam 1949). So, it is definitely Monte Carlo, not Las Vegas.

Random Walk

A random walk is a stochastic process that describes a path made of a series of random steps. Figure 5.2 shows a plot of the positions of eight one-dimensional random walks that start at the origin. On the X axis, we have the number of steps, from 0 to 100. On the Y axis, we have the current position on the line (since it is a one-dimensional walk).

Figure 5.2: Example of Eight Random Walks Starting at the Origin

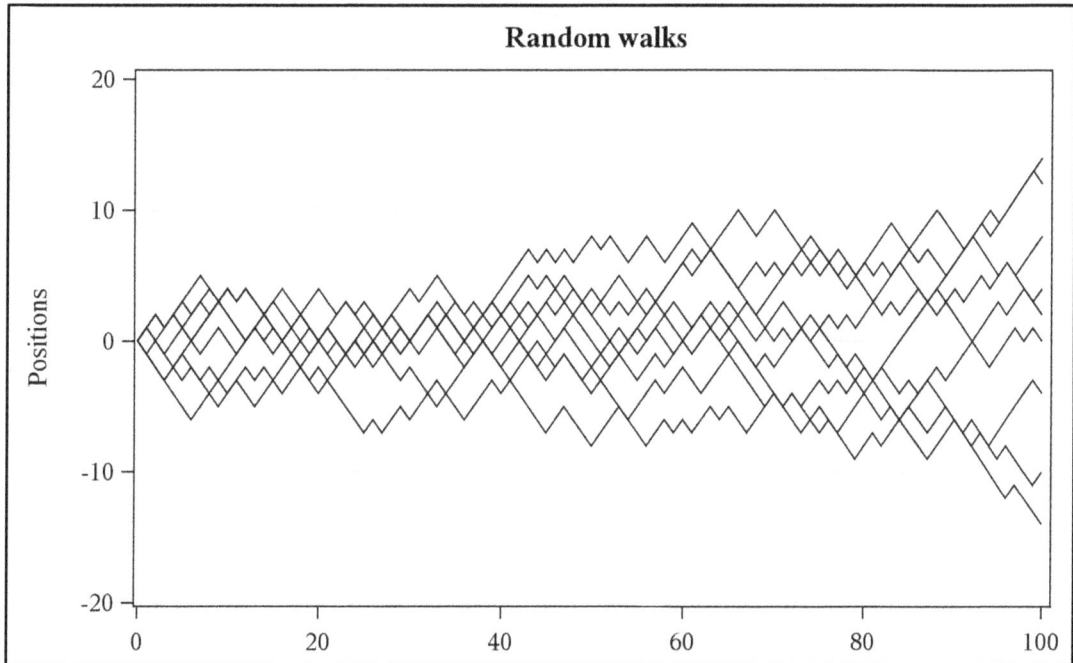

Random walks were initially introduced by Karl Pearson in 1905 (Pearson 1905). There are many applications of random walks in a variety of fields. Here are a few examples:

- In computer science, the page rank algorithm is based on the stationary distribution of a random walk (Brin and Page 1998).
- In finance, the price of stock can be modeled as a random walk (Malkiel 1973).
- In physics, random walks can be used to model the path of a molecule going through a fluid (Bressloff 2014).
- Many more examples can be found at https://en.wikipedia.org/wiki/Random_walk. One of them has an almost comical aspect to it: modeling the ambulation of a drunkard can be done with a random walk and is often used as a didactic tool.

For our purpose, we focus on the simplest possible case, since our aim is to illustrate deep learning for numerical applications (DL4NA), not to study random walks.

So, let's turn our attention to a random walk along an integer line ($\mathbb{Z}$) that starts at 0 and moves +/- 1 with equal probability at each step (for our discussion, we only consider a uniform distribution, but many other distributions could be considered).

This is the random walk that we plotted in Figure 5.2. More specifically, we focus on computing the distance from the origin after n steps.

After 0 steps, the distance from the origin is necessarily 0.

After 1 step, the distance from the origin is necessarily 1, since there are only two possible positions (the random walk as we defined must move +/- 1 and cannot remain stationary):

- -1 if we took a step backward
- +1 if we took a step forward

At step 2 there are a few more possibilities for the position:

- If the position at step 1 is -1:
  - -2 if we took a step backward
  - 0 if we took a step forward
- If the position at step 1 is +1:
  - 0 if we took a step backward
  - 2 if we took a step forward

Consequently, after 2 steps, the distance from the origin can be 0 or 2.

Since this is becoming a little tedious, let's write some code for it.

First, we need to compute the positions at each step:

Program 5.3: Computing Positions at Each Step

```
subroutine move (
  x);     /* IN/OUT The initial (and final) x position */
  outargs x;

  random_number = rand("UNIFORM");
  if random_number <= 0.5 then step = -1; else step = 1;
  x = x + step;  /* Take step in the x direction */
endsub; /* move */
```

As you can see, we rely on the SAS RAND function and the uniform distribution. Given the random value (0 < value < 1), we decide to take a step forward or a step backward.

Now let's compute the distance from the origin after n steps (called `nb_steps` in our program):

Program 5.4: Distance from the Origin after n Steps

```
function take_a_walk(
  nb_steps,    /* IN      The number of steps to take.      */
  x);          /* IN/OUT The initial (and final) x position */
  outargs x;
  do ndx = 0 to nb_steps; /* We walk nb_steps */
    call move(x);
  end;
  return (euclid(x));
endsub; /* take_a_walk */
```

We simply call `move` as many times as we have steps and compute the Euclidian distance from the origin (we could simply take the absolute value, but at this point, we favor readability over performance).

In order to run this code with multiple different values, let's pass the number of steps from the command line:

Program 5.5: Passing the Number of Steps

```
%let nb_steps = %sysget(nb_steps);

data _null_;
  x = 0;
  distance = take_a_walk(&nb_steps., x);
  put "nb_steps: &nb_steps., final position: " x=',
      distance from origin: ' distance=;
run;
```

Now let's run our program:

Program 5.6: Computing Distance from the Origin for a Random Walk

```
$ sas rw.sas -set nb_steps 500000 -stdio
...
1          /**
2             \file
3             \brief This SAS program computes the distance
4                    from the origin for a 1D random walk
5                    (# steps are passed on the command line).
6
7             \author SAS Institute INC.
8             \date 2018
9          */
10         options cmplib = work.functions;
11
12
13         proc FCMP outlib = work.functions.rw;
14            /*
15             Take one step in a 1D random walk.
16            */
17            subroutine move (
18              x);     /* IN/OUT The initial (and final) position */
19              outargs x;
20              random_number = rand("UNIFORM");
21              if random_number <= 0.5 then step = -1;
22              else step = 1;
23              x = x + step;  /* Take step (could be backwards) */
24            endsub; /* move */
25
26            /*
27             Take a random walk of nb_steps from the position
28             passed in. Returns the (Eunclidian) distance
29             from the origin.
30            */
```

```
31              function take_a_walk(
32                nb_steps,  /* IN The number of steps to take.    */
33                x);    /* IN/OUT The initial (and final) position */
34                outargs x;
35                do ndx = 0 to nb_steps; /* We walk nb_steps */
36                  call move(x);
37                end;
38                return (euclid(x));
39              endsub; /* take_a_walk */
40          quit;

NOTE: Function take_a_walk saved to work.functions.rw.
NOTE: Function move saved to work.functions.rw.
NOTE: PROCEDURE FCMP used (Total process time):
      real time            0.02 seconds
      cpu time             0.02 seconds


41
42          %let nb_steps = %sysget(nb_steps);
43
44          data _null_;
45            x = 0;
46            distance = take_a_walk(&nb_steps., x);
47            put "nb_steps: &nb_steps., final position: " x=;
48            put 'distance from origin: ' distance=;
49          run;

nb_steps: 500000, final position: x=513
distance from origin: distance=513
NOTE: DATA statement used (Total process time):
      real time            0.03 seconds
      cpu time             0.04 seconds


50

NOTE: SAS Institute Inc., SAS Campus Drive, Cary, NC USA 27513-2414
NOTE: The SAS System used:
      real time            0.10 seconds
      cpu time             0.08 seconds
```

If we run it ten times, we get very different results each time:

```
distance from origin: distance=277
distance from origin: distance=529
distance from origin: distance=1089
distance from origin: distance=329
distance from origin: distance=221
distance from origin: distance=231
distance from origin: distance=35
distance from origin: distance=169
distance from origin: distance=195
distance from origin: distance=715
```

This is to be expected because of the stochastic nature of our process. The variance of those values is quite large: 40,963. That is not good!

It can be proven that the distance from the origin for large values of n is approximately $\sqrt{2 * n / \pi}$, but in our case we want to showcase Monte Carlo simulations. So, to compute the expected value of a stochastic variable (the distance from the origin), we will resort to a Monte Carlo simulation. Let's do that with a very simple change to our program:

Program 5.7: Using a Monte Carlo Simulation

```
%let nb_steps = %sysget(nb_steps);
%let nb_trials = %sysget(nb_trials);

data _null_;
  accum = 0;
  do ndx_trial = 1 to &nb_trials.;
    x = 0;
    distance = take_a_walk(&nb_steps., x);
    accum = accum + distance;
  end;
  accum = accum / &nb_trials.;
  put "nb_steps: &nb_steps., final position: " x=;
  put 'average distance from origin: ' accum=;
run;
```

As you can see, we simply loop over the random walk function and average the computed distance from the origin. Let's run the program once:

```
$ sas rw_mc.sas -set nb_steps 500000 -set nb_trials 500 -stdio
...
42            %let nb_steps = %sysget(nb_steps);
43            %let nb_trials = %sysget(nb_trials);
44
45            data _null_;
46              accum = 0;
47              do ndx_trial = 1 to &nb_trials.;
48                x = 0;
49                distance = take_a_walk(&nb_steps., x);
50                accum = accum + distance;
51              end;
52              accum = accum / &nb_trials.;
53              put "nb_steps: &nb_steps., final position: " x=;
54              put 'average distance from origin: ' accum=;
55            run;

nb_steps: 500000, final position: x=51
average distance from origin: accum=536.104
NOTE: DATA statement used (Total process time):
      real time         14.14 seconds
      cpu time          14.15 seconds
```

And now ten times:

```
average distance from origin: accum=580.404
average distance from origin: accum=555.58
average distance from origin: accum=628.924
average distance from origin: accum=558.076
average distance from origin: accum=548.008
average distance from origin: accum=552.756
```

```
average distance from origin: accum=567.712
average distance from origin: accum=580.796
average distance from origin: accum=582.616
average distance from origin: accum=590.112
```

This is much better! The variance of those results is much lower than without the simulation; it is now 577. Note that the theoretical value for a large value of n (in this case 500,000) is $\sqrt{2 * n / \pi} \approx 564.19$.

If you run this experiment on your machine, you will see different results, because the series of random numbers will be different and consequently the series of steps will be different.

This improved stability comes at a great cost: the execution time. To compute an approximation of the distance from the origin in a random walk, we now take over 14 seconds:

```
NOTE: DATA statement used (Total process time):
      real time           14.14 seconds
      cpu time            14.15 seconds
```

It was a fraction of a second without the simulation.

We could reduce the number of steps. Let's say 50:

```
average distance from origin: accum=521.84
average distance from origin: accum=512.56
average distance from origin: accum=628.24
average distance from origin: accum=584.68
average distance from origin: accum=618.92
average distance from origin: accum=576.4
average distance from origin: accum=492.32
average distance from origin: accum=562.72
average distance from origin: accum=455.6
average distance from origin: accum=525.04
```

In this case the variance goes up significantly: 3,105. This might not be palatable for most applications. In fact, we might need to increase the number of steps to increase our accuracy. For example, here are the results with 1,000 steps:

```
average distance from origin: accum=550.122
average distance from origin: accum=566.424
average distance from origin: accum=535.296
average distance from origin: accum=545.066
average distance from origin: accum=556.608
average distance from origin: accum=537.464
average distance from origin: accum=552.126
average distance from origin: accum=543.368
average distance from origin: accum=558.84
average distance from origin: accum=563.246
```

The variance now goes down to 112. But the execution time goes up accordingly:

```
NOTE: SAS Institute Inc., SAS Campus Drive, Cary, NC USA 27513-2414
NOTE: The SAS System used:
      real time           29.27 seconds
      cpu time            29.24 seconds
```

If we're considering a stock price simulation with 1,000,000 positions or an n-body simulation, the overall computation time would be exorbitant: 1,000,000 * 30 seconds ≈ 6 days. Please keep in mind that the stochastic process we are considering is quite trivial. A model that prices stocks or complicated insurance contracts like variable annuities would be impractical.

We needn't worry, though, because we have at our disposal many cores that will divide this computation time to something more reasonable.

For the remainder of our random walks, we focus on the following parameters:

- 300K to 500K steps
- 500 trials (sometimes called draws, as in "drawing a card from a deck")
- 10,000 distances to compute (between 300K to 500K steps)

That choice of 10K distances to compute is somewhat arbitrary. That number can be considered very small for applications in physics, small for applications in finance, and very big for modeling a drunkard (Tzou 2014; Komarov and Winkler 2013). But making an arbitrary choice doesn't matter in this case, because we are interested in the relative execution times of the methodologies that we will try. We will experiment with three implementations of a Monte Carlo simulation:

- single-threaded
- multi-threaded
- deep learning (DL)-powered

The single-threaded simulation is the one we just did. We used only one core of the machine for the simulations. We will do the multi-threaded simulation shortly, and the DL-powered version in Chapter 7.

Speaking of hardware, let's say a few words about the machine we are running on. It is called FSNLAX05, as you can infer from some of the screenshots. The machine runs Red Hat LINUX 7.3 and has 64 Hyper-Threaded cores running at 2.3 GHz:

```
$ cat /etc/redhat-release
Red Hat Enterprise Linux Server release 7.3 (Maipo)
$ tail -n 25 /proc/cpuinfo
vendor_id  : GenuineIntel
cpu family : 6
model           : 63
model name : Intel(R) Xeon(R) CPU E5-2698 v3 @ 2.30GHz
stepping    : 2
microcode  : 0x38
cpu MHz     : 2509.335
cache size : 40960 KB
physical id        : 1
siblings    : 32
core id     : 15
cpu cores   : 16
```

You can tell that this is a two-socket machine (two physical Xeon chips) because the last physical ID is 1 (the first is 0). It is not an extremely fast machine, but 64 threads are a decent number (2 * 16 CPU cores * 2 for hyperthreading = 64 threads).

FSNLAX05 also has two NVIDIA GPU devices:

```
$ nvidia-smi
+-----------------------------------------------------------------------------+
| NVIDIA-SMI 375.26                 Driver Version: 375.26                     |
|-------------------------------+----------------------+----------------------+
| GPU  Name        Persistence-M| Bus-Id        Disp.A | Volatile Uncorr. ECC |
| Fan  Temp  Perf  Pwr:Usage/Cap|         Memory-Usage | GPU-Util  Compute M. |
|===============================+======================+======================|
|   0  Tesla K80           Off  | 0000:05:00.0     Off |                    0 |
| N/A   35C    P8    27W / 149W |      2MiB / 11439MiB |      0%      Default |
+-------------------------------+----------------------+----------------------+
|   1  Tesla K80           Off  | 0000:06:00.0     Off |                    0 |
| N/A   29C    P8    30W / 149W |      2MiB / 11439MiB |      0%      Default |
+-------------------------------+----------------------+----------------------+
```

We will make extensive use of those GPU devices during input data generation, deep neural network training, and deep neural inference (what we used to call scoring before DL). The 0% means that the devices are not busy, but we will change that!

Now that we have seen that 10,000 measurements of the expected distance would take approximately 10,000 * 14.14 seconds = 141,400 seconds ≈ 39 hours, let's see how much better we can do with more threads.

Multi-threaded Random Walk

SAS Studio

As you have undoubtedly noticed, we have been using the command line version of SAS (stdio option). We could stick to the command line version for our multi-threaded work, but switching to a more GUI-oriented environment allows us to better visualize the execution on many threads, so let's do that.

When you start SAS Studio and enter your credentials, assuming that you have set up a link as indicated in Appendix A, your screen should look something like this:

Figure 5.3: SAS Studio

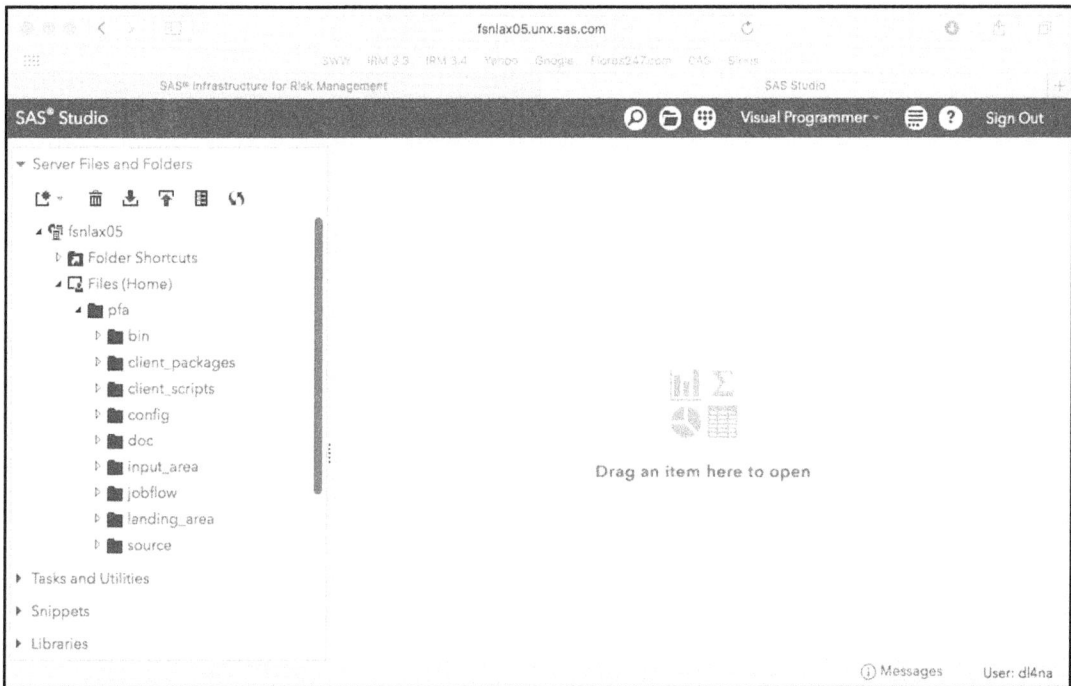

Our goal at this point is to create a SAS Infrastructure for Risk Management task to perform one random walk for a given number of steps. So let's copy and paste the previous code and save it to the `nodes` folder (See Chapter 4 on many-task computing for further explanations):

Figure 5.4: SAS Studio with Random Walk Task

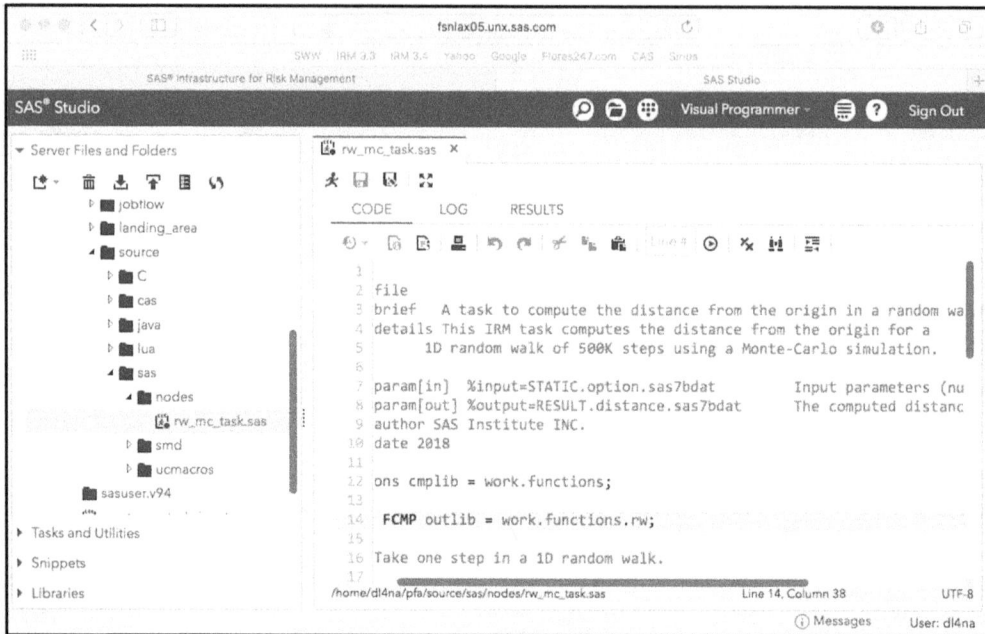

As we discussed in Chapter 4, a task *must* declare its inputs and outputs, and those inputs *must* come from a file, typically a SAS data set. Consequently, we need to modify the following lines of code and put the equivalent in an input data set:

```
%let nb_steps = %sysget(nb_steps);
%let nb_trials = %sysget(nb_trials);
```

Since we're SAS developers, that is no problem. First we write a couple of lines of SAS code to create a data set with those values:

Program 5.8: Creating an Input Data Set

```
libname ia '~/pfa/input_area';
data ia.option;
  input nb_trials nb_steps;
  datalines;
500 500000
;

data _null_ ;
    FILE "~/pfa/config/libnames.txt";
    PUT 'STATIC=%la';
run ;

/* We also need to touch last_update.txt to trigger live ETL and a refresh of the
libnames */
data _null_ ;
    FILE "~/pfa/input_area/last_update.txt";
    PUT;
run ;
```

Live ETL

You'll notice that we've put those data sets in the `input_area` folder. This requires a little bit of explanation. All inputs in SAS Infrastructure for Risk Management come from a folder called the `landing_area` (`~/pfa/landing_area`), so this is ultimately where we want our input file. However, SAS Infrastructure for Risk Management can run many flows and many tasks at the same time, so if a task happens to be running while you're copying data into `landing_area`, then things will not go smoothly. You might get errors because the file is locked or even incorrect results if more than one input needed to be copied. For this reason, it is best to let SAS Infrastructure for Risk Management copy the data from the `input_area` folder to the `landing_area` folder when it knows for sure that it is safe to do so. More details can be found in the *SAS Infrastructure for Risk Management: Programmer's Guide* (available at http://go.documentation.sas.com/api/docsets/irmscug/3.4/content/irmscug.pdf).

The last two DATA steps are required for the following reasons:

1. Since SAS Infrastructure for Risk Management owns the libref assignments, we need to put enough information for our libref assignments into libname.txt. The `%la` refers to the `landing_area` folder; this is convenient to avoid full paths that won't work when the code is moved to another machine.

2. Since we need SAS Infrastructure for Risk Management to copy the files from the `input_area` folder to the `landing_area` folder, we need to tell SAS Infrastructure for Risk Management when we're done populating the `input_area` folder. Touching `last_update.txt` will trigger the copy from `input_area` to `landing_area`.

After execution, you should see the following in SAS Studio:

Figure 5.5: SAS Studio with Random Walk Task Input

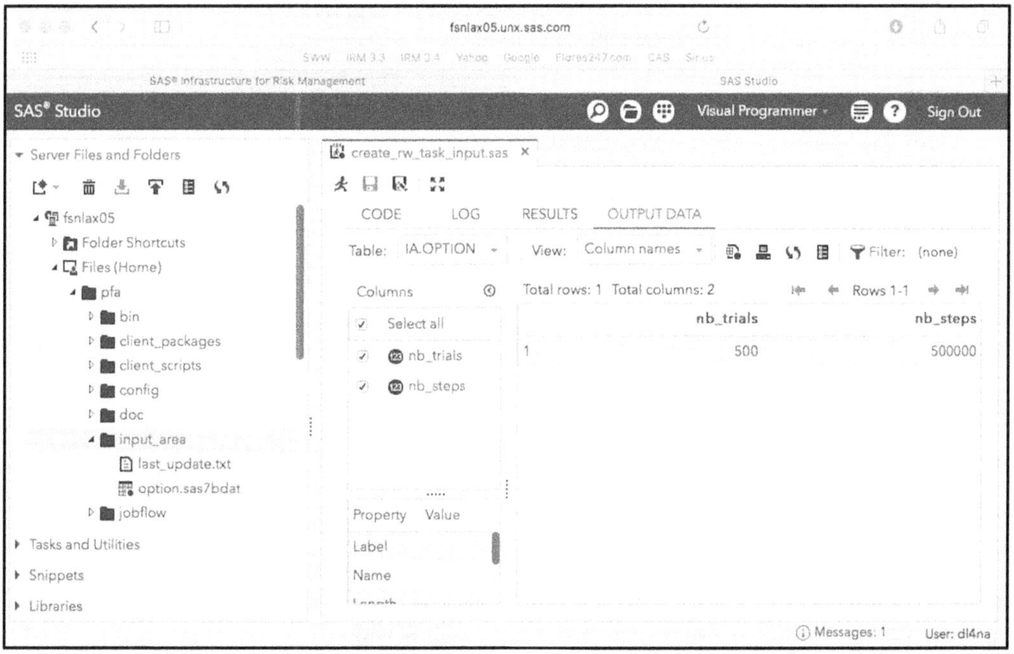

And after a short delay (up to 30 seconds), `landing_area` is populated as required:

Figure 5.6: SAS Studio with Random Walk Task Input after Live ETL.

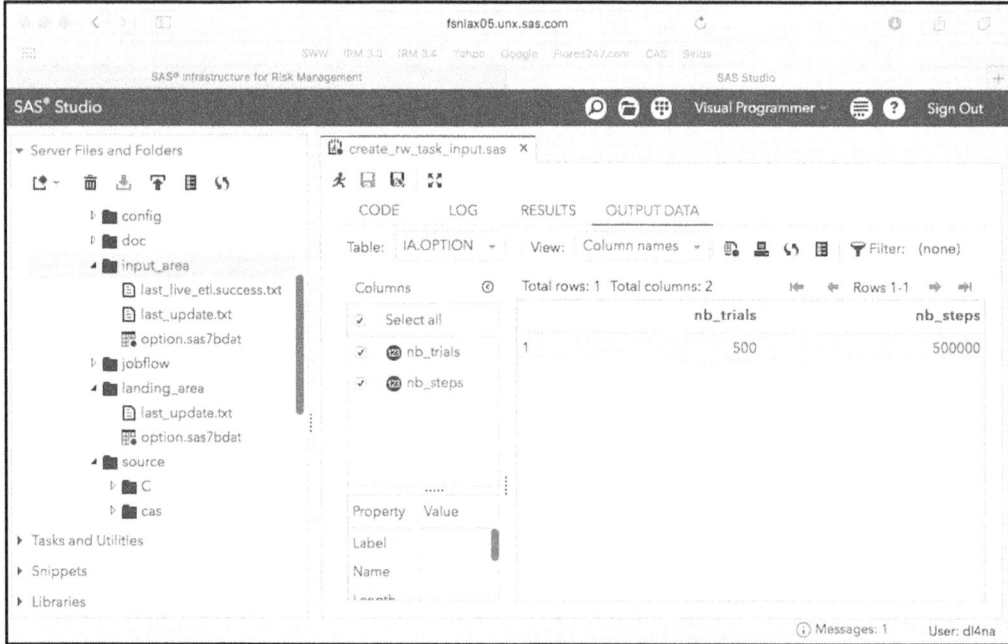

You'll notice that SAS Infrastructure for Risk Management added an extra file (`last_live_etl_success.txt`) to notify you of success or failure. In our case, it looks like success (`last_live_etl_error.txt` would indicate failure).

A Parallel Program

Next, we modify the header of our task to define the new input and output:

Program 5.9: Defining the New Input and Output

```
/**
   \file
   \brief   A task to compute the distance from the origin in a random walk.
   \details This SAS Infrastructure for Risk Management task computes
            the distance from the origin for a 1D random walk of
            500K steps using a Monte Carlo simulation.

   \param[in]  %input=STATIC.option.sas7bdat     Input parameters (number of
                                                  steps and trials).
   \param[out] %output=RESULT.distance.sas7bdat  The computed distance.
   \author SAS Institute INC.
   \date 2018
*/
```

Now we need to modify our task in order to generate the output that we claim to generate (in our command-line version we merely printed the results via a PUT statement):

Program 5.10: Generating Output

```
data RESULT.distance(keep=average);
  set STATIC.option;
  accum = 0;
  do ndx_trial = 1 to nb_trials;
    x = 0;
    distance = take_a_walk(nb_steps, x);
    accum = accum + distance;
  end;
  average = accum / nb_trials;
  put "nb_steps: nb_steps, final position: " x=;
  put 'average distance from origin: ' average=;
run;
```

The changes are in **bold**. You'll notice that we now read the options (number of steps and number of trials) from the STATIC.option SAS data set, which contains only one row at this point. This means that we will have only one set of draws for a 500K random walk.

That's almost it. We have a task, we have inputs, but we don't have a parallel program. So let's create a parallel program. We will call it rw_mc_flow.sas. A few lines of SAS code will suffice:

```
%irm_sc_init();
```

The preceding line is required to initialize the scripting client. It mostly defines the macros that we will use, but also defines the connection parameters to the SAS Infrastructure for Risk Management server and some debug options. Here, we rely on the default (the SAS Infrastructure for Risk Management server is on the same machine as the SAS Studio server), so we don't pass any additional parameters.

Next, each SAS Infrastructure for Risk Management parallel program or job flow definition starts with the definition of a job flow:

```
%irm_sc_build_jobflow(
    i_jf_name   =random_walk_mc,
    o_jf_ref_name =random_walk_mc_ref);
```

At this point, our job flow is called random_walk_mc and we use random_walk_mc_ref as a reference inside this SAS code (for example, to execute the parallel program or an instance of the parallel program, to be more exact). Now, that we have a job flow, we must add a task to it to do something useful:

```
%rw_mc_task(
    i_jf_ref =&random_walk_mc_ref);
```

Note that it is not valid to create an empty job flow (with no task).

As we discussed in Chapter 4, SAS Infrastructure for Risk Management has automatically created a macro by the same name as our task so that we can easily add a task to a parallel program (our task is in rw_mc_task.sas). Consequently, the preceding statement adds rw_mc_task to our job flow with the defaults inputs and output. At this point, we can simply save and execute the definition of our parallel program:

Program 5.11: Parallel Program Definition

```
%irm_sc_save_jobflow(
    i_jf_ref =&random_walk_mc_ref);

%irm_sc_execute_jobflow(
    i_jf_ref =&random_walk_mc_ref);
```

If you run this code in SAS Studio, you should see something like this indicating a successful run:

Figure 5.7: Successful Run in SAS Studio.

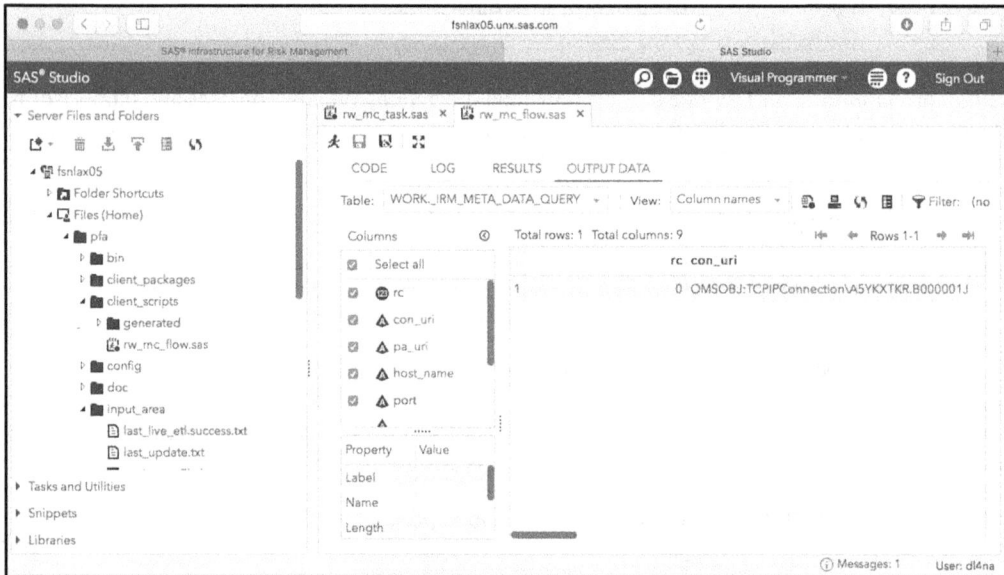

We can go to SAS Infrastructure for Risk Management to see the results of the execution of the job flow instance (as opposed to the results of the execution of the creation of the job flow definition, which is what we see in SAS Studio in the preceding figure):

Figure 5.8: Successful Run in SAS Infrastructure for Risk Management.

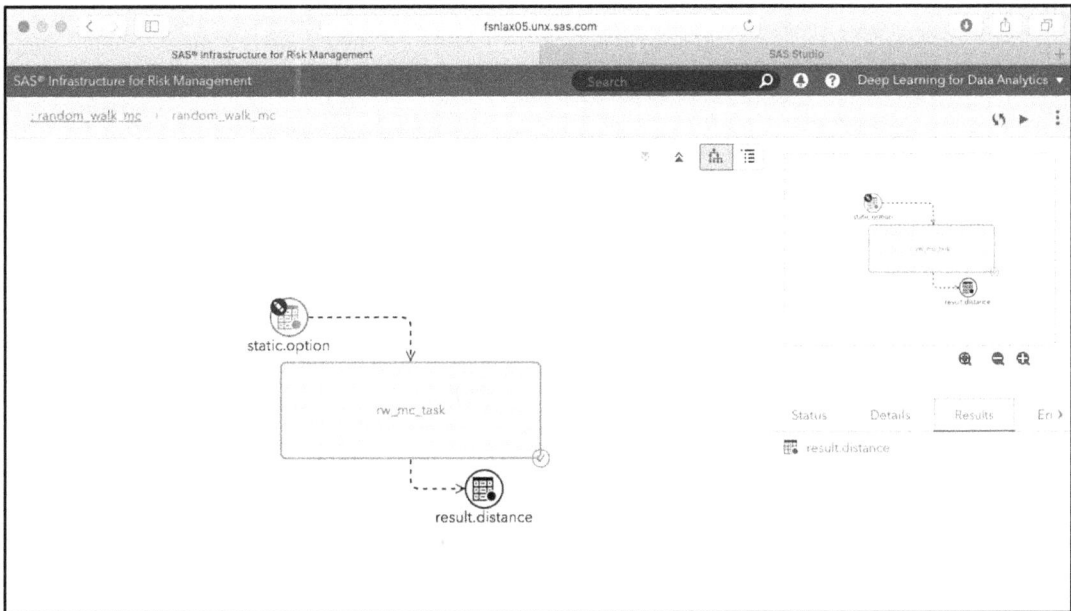

If you double-click on the `result.distance` data object, SAS Infrastructure for Risk Management downloads its content and loads it into an Excel spreadsheet:

Figure 5.9: RESULT.distance in Excel.

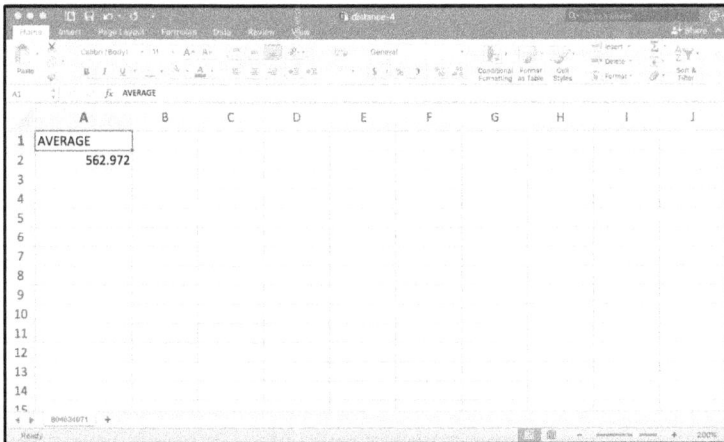

Depending on your browser and its settings, the file might open directly or be available from your download folder. Pay special attention to pop-up blockers that might prevent you from downloading and opening the file.

As you might expect, if you go through the same exercise on your machine, then your average is likely to be different. Alternatively, you can also view the table in SAS Studio via the pop-up menu for `result.distance`.

The SAS Infrastructure for Risk Management GUI allows you to view the log for the task by using the pop-up menu:

Figure 5.10: SAS Infrastructure for Risk Management GUI Pop-Up Menu.

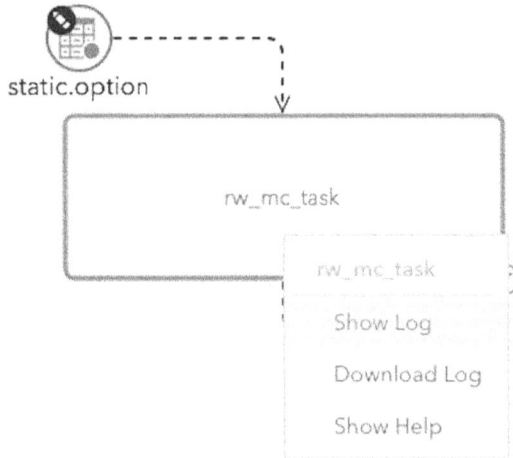

The log shows that we took about 15 seconds:

```
nb_steps: nb_steps, final position: X=-1497
average distance from origin: AVERAGE=575.064
NOTE: There were 1 observations read from the data set STATIC.OPTION.
NOTE: The data set RESULT.DISTANCE has 1 observations and 1 variables.
NOTE: DATA statement used (Total process time):
      real time           15.48 seconds
      cpu time            15.47 seconds
```

This result is not surprising, since it is similar to the 14 seconds that we observed from the command line, plus some overhead that is probably due to the fact we now open a SAS data set to get our inputs.

A Parallel Program with Partitions

As far as parallel programs go, our example is limited. We have only one task, it produces one result, and we run on only one thread. Let's see what we can do to change that.

The first thing to do is to expand our input data (STATIC.option) so that we take 10,000 measurements instead of one:

Program 5.12: Expanding the Input Data

```
data ia.option(keep=nb_trials nb_steps);
  do ndx = 0 to 10000;
    nb_trials = 500;
    nb_steps  = int(300000 + rand("UNIFORM") * 200000);
    output;
  end;
  stop;
run;
```

If we were to go at it single-threaded, this would take us about 150,000 seconds, or a little less than 42 hours. Let's not do that! Instead, let's modify our flow creation to include partitions so that we can maximize the usage of our hardware. SAS Infrastructure for Risk Management greatly facilitates this process, as you will see in a few moments.

We go back to rw_mc_flow.sas, which contains the definition of our parallel program, and add the following two macro calls before and after the task invocation (rw_mc_task):

Program 5.13: Adding Partitions

```
%irm_sc_add_partition_byn_task(
        i_jf_ref           =&random_walk_mc_ref,
        i_task_name        =partition_option_byn,
        i_card_tb_byn      =STATIC.cardinality.sas7bdat,
        i_part_ds_byn      =STATIC.option.sas7bdat,
        i_out_part_tb_byn =P_STAGE.option.sas7bdat);

%rw_mc_task(
    i_jf_ref =&random_walk_mc_ref,
        t_input  =P_STAGE.option.sas7bdat,
        t_output =P_STAGE.distance.sas7bdat);

%irm_sc_add_recombine_byn_task(
        i_jf_ref           =&random_walk_mc_ref,
        i_task_name        =recombine_distance_byn,
        i_rec_tb_in_byn   =P_STAGE.distance.sas7bdat,
        i_rec_out_tb_byn  =SIMUL.average_distances.sas7bdat);
```

The irm_sc_add_partition_byn_task macro creates a partitioned version of STATIC.option.sas7bdat where the number of partitions is given by STATIC.cardinality.sas7bdat. The partitioned version of the option table goes into P_STAGE.option.sas7bdat. If you were to look on the file system, you would see 60 different option tables (we use 60 threads to leave some room for the GUI and the SAS Infrastructure for Risk Management middle tier):

Figure 5.11: Partitioned Option Data Set.

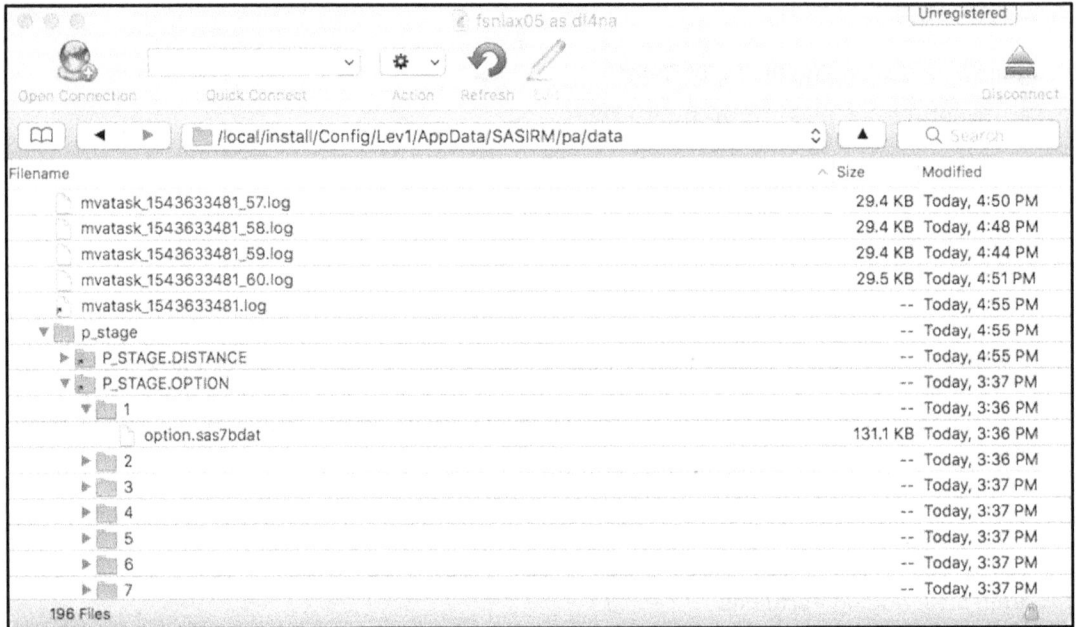

...

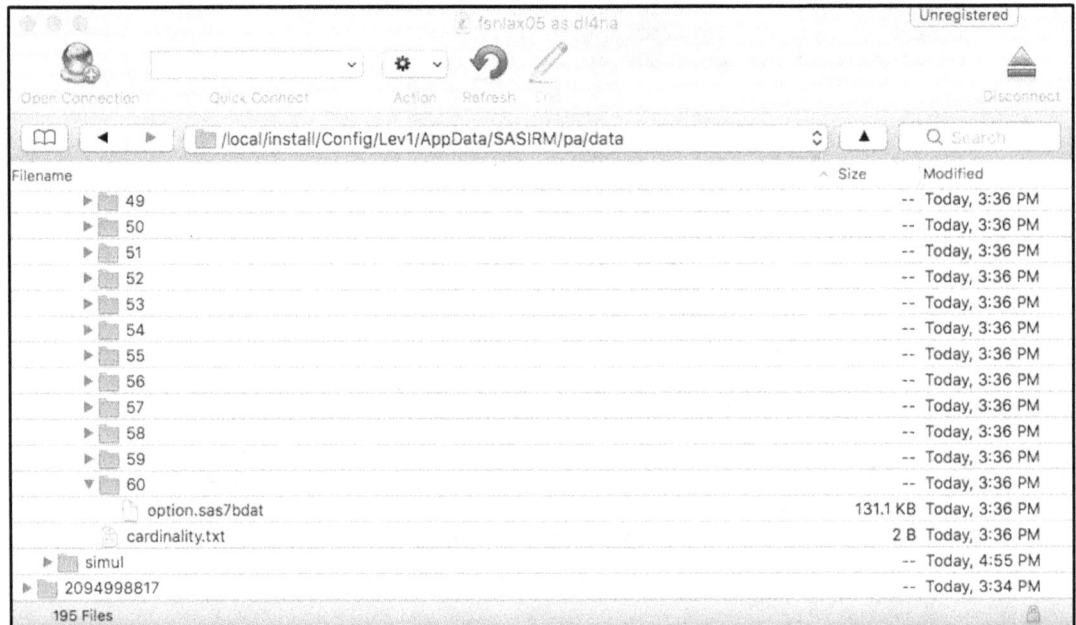

Each of these tables is used by one of the 60 parallel executions of the task. Another way to present what is going on is to state that the partitioned version of the option table is 60 different option tables, each with its own content: one row with the NB_STEPS and NB_TRIALS variables. (All rows have the same value for

NB_TRIALS (500), but a different value for NB_STEPS, between 300K and 500K, as we discussed earlier.)

Since the partitioned version of the option table goes into the `p_stage` library, we can no longer rely on the default arguments of the task. The `p_stage` library is created in the persistent area (PA) of SAS Infrastructure for Risk Management, not in the **landing_area** folder, since it is an output. The **landing_area** is reserved for inputs of parallel programs.

The `irm_sc_add_recombine_byn_task` macro does the converse of the partitioning macro: it recombines the partitioned outputs into one table, namely SIMUL.average_distances.sas7bdat.

With the addition of those two macros, when the `rw_mc_task` executes, it will in fact run 60 separate processes and consequently go from a single-threaded program to a multi-threaded one. There are a lot of orchestrations and synchronizations behind the scenes, but SAS Infrastructure for Risk Management makes this process semi-transparent to the SAS programmers after the partitioning and recombining macros are added. This high level of automation of the tedious and error-prone multi-threading development task is one the main reasons to use SAS Infrastructure for Risk Management. The other key reason is integration, as we will discuss in Chapter 8, "Deep Learning for Numerical Applications in the Enterprise."

We now have a parallel program that can run on hundreds of cores. Well, almost. We need to slightly modify our task in order to keep the number of steps that we simulated:

```
data &output.(keep=average nb_steps);
  set &input.;
  accum = 0;
  do ndx_trial = 1 to nb_trials;
```

We are now ready to create our new parallel program and run it. So let's do that and run it in SAS Studio as we did before. The net result will be a partitioned flow (parallel program) like the one in Figure 5.12.

Figure 5.12: A Partitioned Flow

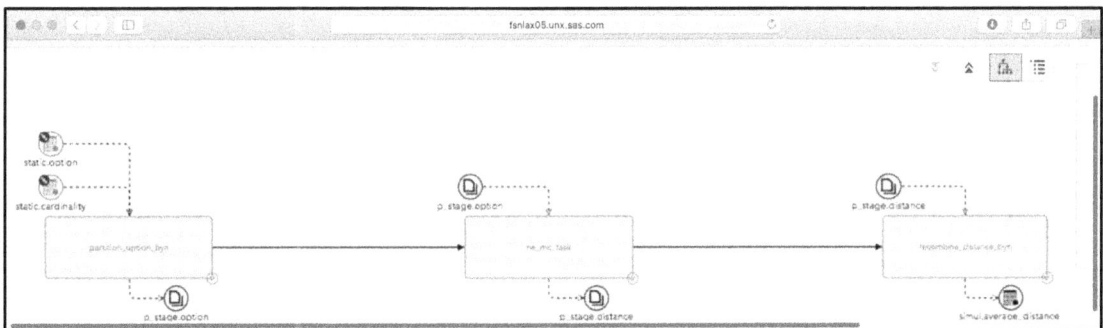

When the Monte Carlo simulation is running, we can see that we indeed use the machine at close to its maximum potential (64 threads), as you can see in Figure 5.13. There are a few threads left so that the GUI can be responsive while the Monte Carlo simulation is running.

Note that to capture the activity on the machine, we used **N**igel's Performance **Mon**itor (NMON), a very valuable tool when it comes to quantifying hardware utilization. NMON can be downloaded from http://nmon.sourceforge.net.

Figure 5.13: NMON with 64 threads on a 64-core machine

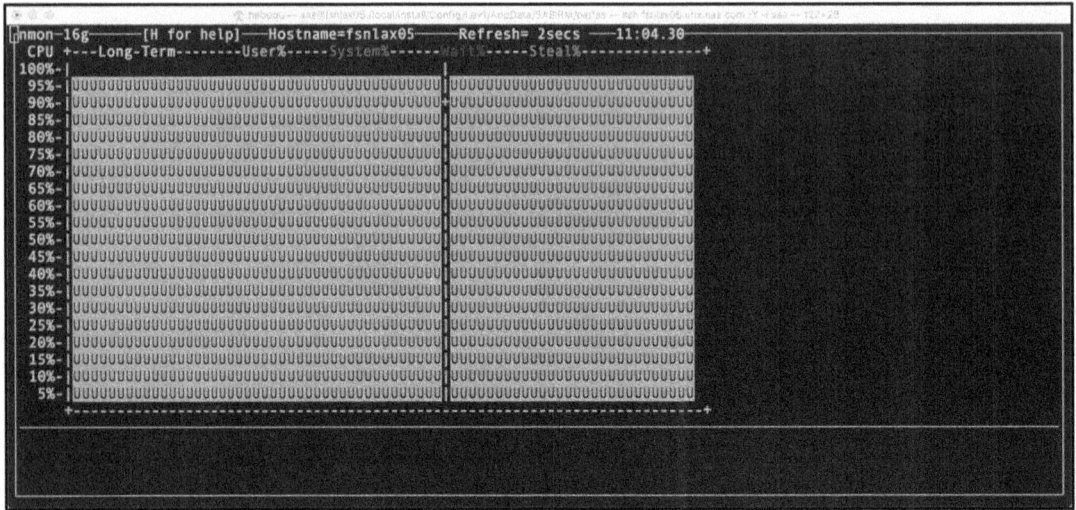

Figure 5.14 shows the results of our simulation: 10,000 average distances for 10,000 random walks (each walk has a potentially different number of steps):

Figure 5.14: Simulation Results

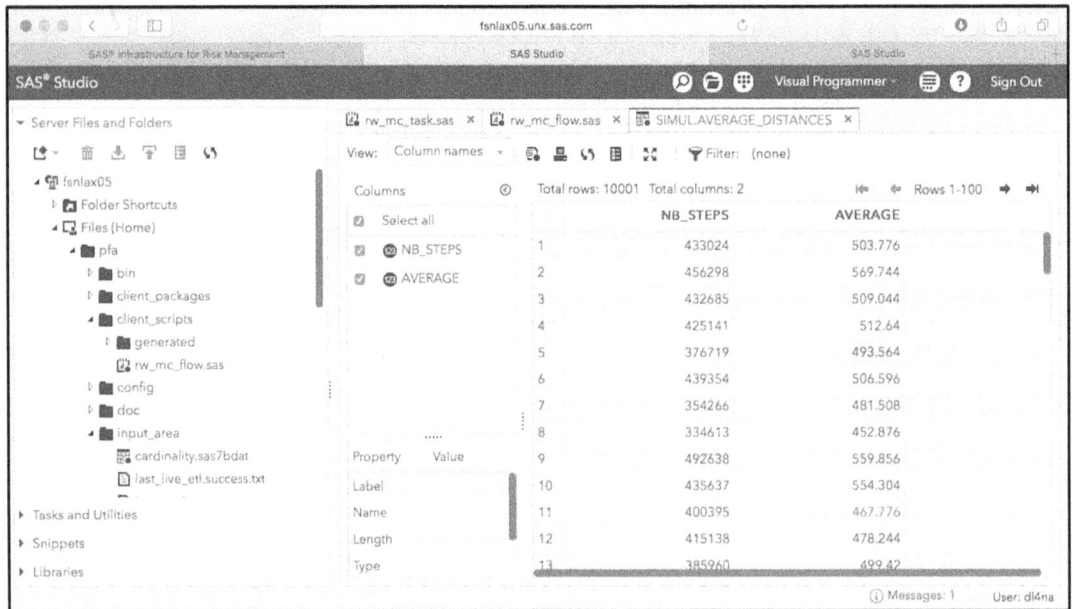

The total execution time of the simulation is 4,393 seconds (1 hour, 13 minutes and 13 seconds). That is indeed much better than the 42 hours that we were contemplating with a single-threaded program.

If you run the experiment on your machine, you might notice that towards the end of the 73 minutes, the activity of the CPU tapers off. Indeed, a close examination of the SAS Infrastructure for Risk Management log reveals that the first simulation stops at around 12:58, but it takes over 8 minutes for the last task and consequently the whole job flow instance to finish (the relevant times are in **bold**):

```
IRM 12:51:16 Done running rw_mc_task (partition rank=59, execution time 3,982,161ms).
...
IRM 12:57:55 Done running rw_mc_task (partition rank=26, execution time 4,381,093ms).
IRM 12:57:55 The job flow random_walk_mc has completed in 4393642 ms.
```

That delay is unfortunate, because we leave about 10% of our final running time on the table. We could redesign the experiment to focus on individual draws rather than 500 draws for a given number of steps, which would complicate things a bit. Alternatively, we can wait until we use a deep neural network to compute our estimates, and we'll see that this problem has disappeared. So let's wait until Chapter 7.

To close this chapter, we should also point out that thanks to SAS Infrastructure for Risk Management, the SAS programmer gets performance portability. Our parallel program works on 64 cores, but could as easily work on a grid or on a machine with hundreds of cores. So why don't we try it?

Many Cores

Let's log in to a machine with 208 threads of execution:

Figure 5.15: A Machine with 208 Threads

```
sas@skylake4:~ > tail -n 27 /proc/cpuinfo

processor       : 207
vendor_id       : GenuineIntel
cpu family      : 6
model           : 85
model name      : Intel(R) Xeon(R) Platinum 8164 CPU @ 2.00GHz
stepping        : 4
microcode       : 0x200000f
cpu MHz         : 1013.750
cache size      : 36608 KB
physical id     : 3
siblings        : 52
core id         : 29
cpu cores       : 26
apicid          : 251
initial apicid  : 251
fpu             : yes
fpu_exception   : yes
cpuid level     : 22
wp              : yes
flags           : fpu vme de pse tsc msr pae mce cx8 apic sep mtrr pge mca cmov pat pse36 clflush dts acpi mmx fxsr sse sse2 ss ht tm pbe syscall nx pdpe1g
b rdtscp lm constant_tsc art arch_perfmon pebs bts rep_good nopl xtopology nonstop_tsc aperfmperf eagerfpu pni pclmulqdq dtes64 monitor ds_cpl vmx smx est
tm2 ssse3 fma cx16 xtpr pdcm pcid dca sse4_1 sse4_2 x2apic movbe popcnt tsc_deadline_timer aes xsave avx f16c rdrand lahf_lm abm 3dnowprefetch ida arat epb
 pln pts dtherm hwp hwp_act_window hwp_epp hwp_pkg_req intel_pt tpr_shadow vnmi flexpriority ept vpid fsgsbase tsc_adjust bmi1 hle avx2 smep bmi2 erms invp
cid rtm cqm mpx avx512f avx512dq rdseed adx smap clflushopt clwb avx512cd avx512bw avx512vl xsaveopt xsavec xgetbv1 cqm_llc cqm_occup_llc cqm_mbm_total cqm
_mbm_local
bogomips        : 4005.20
clflush size    : 64
cache_alignment : 64
address sizes   : 46 bits physical, 48 bits virtual
power management:

sas@skylake4:~ >
```

When we run the flow this time, the total execution time goes down to 1,353 seconds (down from 4,393):

```
IRM 11:41:33,086 The job flow instance random_walk_mc has completed in 1353326 ms.
```

In other words, we went from over 40 hours down to 23 minutes. Not bad! You might object that you don't have a 200+ core server at your disposal. That might be true, but the important point is that those machines are coming and they will come down in price. It should also be pointed out that this is not the machine with the most threads of execution that money can buy. As of this writing, one can get an 8-socket machine with over 400 threads of executions. Even more important is that these 23 minutes are far from the limit in terms of performance and price/performance ratio.

You might wonder why the progression from 64 threads to 200 is not more linear. The explanation is simple: on the 64-thread machine, the clock is running at 2.3 GHz:

```
model name : Intel(R) Xeon(R) CPU E5-2698 v3 @ 2.30GHz
```

But on the 200-thread machine, the clock is running at 2.00 GHz:

```
model name : Intel(R) Xeon(R) Platinum 8164 CPU @ 2.00GHz
```

This simply translates to more instructions per second per thread (per core) on the 64-thread machine. Note that this is pretty typical. More cores usually imply slower clock rates, unless you go up in machine class.

The number of partitions was set to 200 by changing the cardinality table. Alternatively, we could run a SAS program that queries the system to find out how many cores are available.

In Figure 5.16, you can see that we indeed make very good use of the hardware. SAS Infrastructure for Risk Management clearly delivers on the promise of performance portability: you write a SAS program once without any synchronization or reference to threads, and no matter where you run your code, it will exploit as many cores as are available on the hardware.

Figure 5.16: NMON with 200 Threads

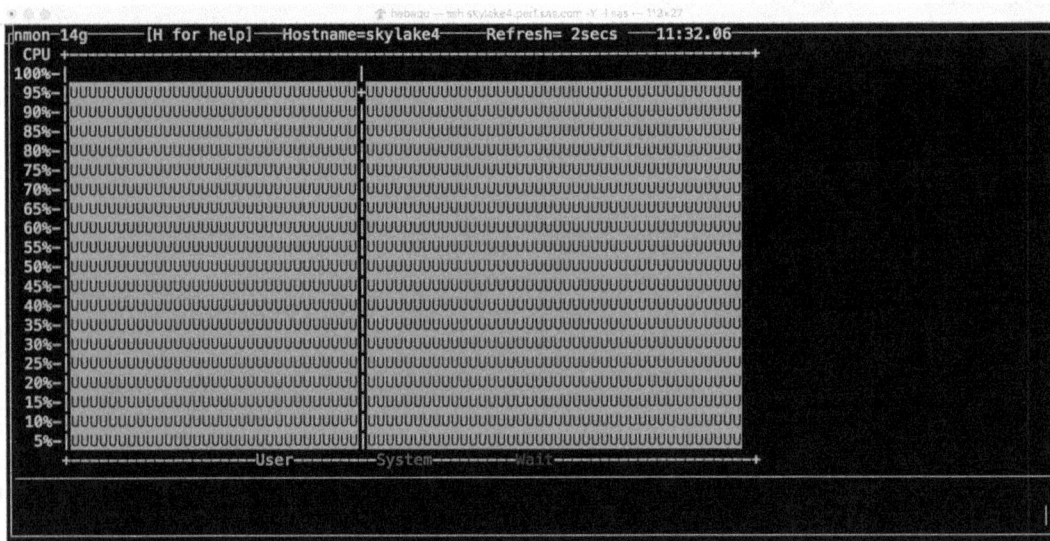

Our performance numbers are summarized in Table 1. Those numbers are impressive on their own, but the best is yet to come. We will dwarf those numbers when we leverage the awesome power of DL with SAS. However, to get there we have to wait until Chapter 7, because we need lots of data to train our neural network. That is the topic of Chapter 6.

Table 5.1: Random Walk Performance

Technology	Execution Time in Seconds
One Thread	150,000
64 Threads	4,393
200 Threads	1,353

Conclusion

In this chapter, we introduced the concept of Monte Carlo simulations, which dates as far back as the 18th century, when Laplace was alive. We then introduced the problem that we will use for the rest of this book as an illustration of a compute-intensive problem: the random walk.

We first wrote a simple SAS program that ran a random walk simulation and realized that to run at scale, this methodology wouldn't quite deliver for all use cases. We then leveraged SAS Infrastructure for Risk Management to run on many core machines: first on a 64-core machine and then on a 208-core machine. The performance gains were significant (over 10x), especially considering the few code changes that we had to make.

Finally, we realized that the load of our random walks was not evenly distributed, which resulted in a sub-optimal use of the hardware that we will tackle in the coming chapters.

In the next chapter, we take a slight detour to have a closer look at a very powerful tool that must be in every statistician's or data scientist's toolbox: GPUs. After the chapter on GPUs, we will resume our work with Monte Carlo simulations.

Chapter 6: GPU

In this chapter, we look at graphics processing units (GPU). We start with a brief history of GPUs, putting the emphasis on the difference between CPUs and GPUs. We then introduce the unique GPU programming model that we leverage to generate training data for the random walk problem that we investigated in the previous chapter.

The goal of this chapter is to briefly introduce GPUs and give you a glimpse of their fast performance and the complexity of their programming. This chapter should be a good motivation for Chapter 7, where we use deep learning (DL) to bypass the complexity of GPU programming, while still reaping all the performance benefits.

If you are already familiar with GPUs and their complex programming model, or if you're ready to start reading about DL, then feel free to skip this chapter altogether.

In this chapter, we build upon the concepts and examples introduced in Chapter 5, so it is a good idea to read Chapter 5 before reading this one.

History of GPUs

In this section, we introduce the problems that motivated the invention of the GPU.

The Golden Age of the Multicore

As discussed in the introduction and in the history of DL in Chapter 2, CPU cores are not getting any faster, because their clock speed has hit a plateau. This wall of performance has motivated chip manufacturers such as Intel and AMD to produce CPUs with more cores. Typically, "more cores" has meant 2 to 32 for CPU manufacturers. Having that many cores motivated some new software designs (or redesigns) for the better part of a decade (from approximately 2005 to 2015). On a single machine, these new designs usually gave us one order of magnitude of performance improvements.

But what if one order of magnitude is not enough? Could we have 2,000 cores and get 3 orders of magnitude of performance improvements? The answer is yes, and the reason is graphics cards.

The Golden Age of the Graphics Card

In parallel to the evolution of the chip manufacturers from single cores to multicores, the graphics industry in general, and the gaming industry in particular, were going through their own evolution. For the graphics industry, the problem in the early 1990s was that CPUs were inherently slow for graphics processing, specifically for visualization and for games.

For applications such as business intelligence (graphs), mapping, and computer-aided design (CAD), the displays of graphs, maps, and drawings was not fast enough for users to be productive. Either you had to sacrifice the quality of your visualization or you had to wait. Lower quality is not an option for applications such as mapping: you need to see the details of the maps that you're working with. Waiting is not an option for applications such as CAD: you must evaluate a lot of designs to get the best possible one.

For 3-D rendering of scenes like the ones that you find in gaming applications, speed is paramount to guarantee realism and immersion of the player into the game.

Why are CPUs inherently bad at solving those problems? Because graphics processing is intrinsically heavily parallel, up to $O(pixels)$.

To satisfy this hunger for speed, companies started offering graphics cards that supported a high level of parallelism through hardwired, pre-implemented logic. Those graphics cards, or early incarnations of the GPU, were designed and built to do one thing: fast graphics processing. They had no programmability and consequently would not meet the definition of what we call a GPU today.

In the mid-1990s to the mid-2000s, there were quite a few successful players. However, as the market matured, most of them got out of the business early (for example, Intergraph), some of them got acquired (for example, AMD acquired ATI in 2006), and only one of them was left standing: NVIDIA.

Note that when we refer to a GPU, we interchangeably mean the card where the GPU is located and the GPU itself (a silicon chip). Most of the time, the difference doesn't matter that much, but it is important to remember that the GPU is a processor like the CPU and it is located on its own board, also known as the GPU card. The GPU card typically has its own private memory, in addition to the CPU memory.

The Golden Age of the GPU

Although courts have stated clearly that NVIDIA didn't invent the GPU, it is true that NVIDIA launched the world's first GPU is 1999 with the GeForce256 (NVIDIA 1999). What distinguished the GeForce256 from its predecessors was its programmability. At the time, the programmability was limited to graphical processing. We would have to wait until 2006 to extend the parallel computational power of GPUs to general programming with the release of CUDA.

With the programmability of the GPU came more flexibility and applications beyond the realm of graphics processing. For this reason, a GPU that is not strictly used for graphics is sometimes called a general-purpose GPU or GPGPU. For most of what we do with GPUs in the context of this book, we need a GPGPU. Since there is no ambiguity in our case, we typically use the acronym GPU rather than the longer GPGPU.

As we discussed in the introduction of the book, GPUs are single instruction stream, multiple data stream (SIMD) processors: they excel at running the same algorithm on different data. To have a better idea of why this is the case, we can look at the diagram in Figure 6.1. The figure shows a streaming multiprocessor (SM), which is at the heart of the computational capabilities of an NVIDIA GPU. One of the most powerful GPUs at the time of this writing, the Volta of Figure 6.1, contains 84 SMs.

If you look at the SM closely, you'll notice that it contains 4 blocks. Each block represents a unit of scheduling in a GPU: a warp. A warp schedules 32 threads that all execute the exact same instruction with each clock cycle. That means that if one of those threads of execution diverges from the others with a test, for example, then all 31 other threads must wait for that single thread to rejoin the group before proceeding. This is the essence of a SIMD processor.

As you can imagine, if your code follows that SIMD model, you get the best possible performance: all threads are busy doing some productive work, and none of them is waiting pointlessly. However, if even a single thread of execution follows its own algorithm (executes its own instruction different from the others), you get worse performance. In that latter case, threads are said to diverge. In GPU programming, you must avoid this divergence of threads.

Here are a few practical examples:

- If you multiply a matrix by a number, the GPU will beat the CPU by orders of magnitude.
- If you sort with a specialized algorithm that makes sure that threads don't diverge, the GPU will beat the CPU.
- If you compute a spreadsheet in which each cell has its own formula, the CPU will beat the GPU by orders of magnitude.

Fortunately for us, machine learning (ML) algorithms are SIMD problems, and a SIMD processor is the best possible processor for them.

Figure 6.1: Volta GV100 Streaming Multiprocessor (Source: Courtesy of NVIDIA)

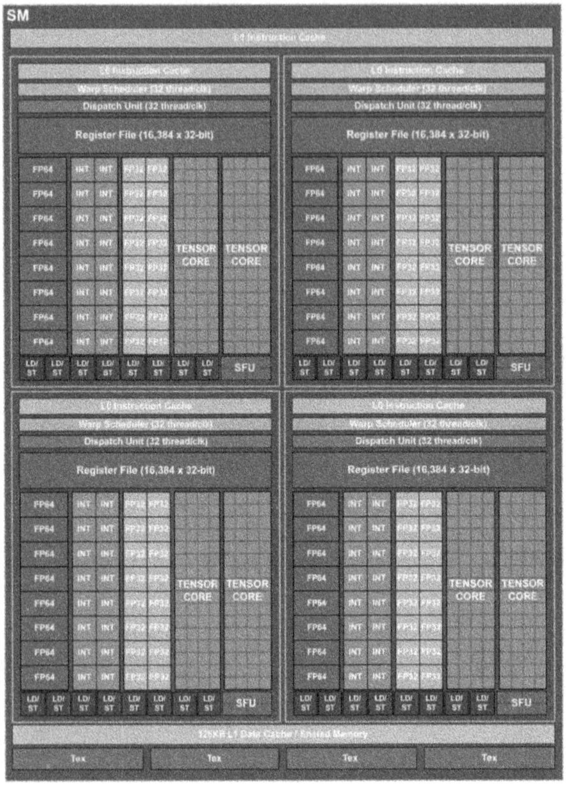

If we examine closely the bottom of Figure 6.1, we notice a few blocks labeled Tex. These are texture mapping units and are a leftover from the GPU graphics processing root.

As a testament to the future of the GPU as opposed to its past, if we look inside each block handled by the warp scheduler, we notice that the SM contains tensor cores (a tensor is a multi-dimensional array). There are 672 tensor cores in the Volta. The presence of the texture mapping units and the tensor cores brings up two observations.

First, a GPU is not a perfect SIMD processor, because it dedicates some silicon to operations that are not general-purpose operations. However, the GPU is the best SIMD processor that money can buy at the time of this writing. We should point out that we qualify this statement in Chapter 9 when we talk about FPGAs.

Second, the design of the GPU is evolving or branching into two designs: a graphics-oriented design for the GPU and a DL-oriented design for the GPGPU.

These two observations show us a likely evolution path for the GPU. At least for DL, the GPU is likely to evolve into an application-specific integrated circuit (ASIC). In this case, the application is DL.

The CUDA Programming Model

Accelerators are nothing new to the computer industry. For example, CPUs typically depend on I/O coprocessors to handle I/O computations. Most accelerators or coprocessors are transparent. Everything works fine with only the CPU, and everything works much faster with the accelerator installed. The GPU is different. Unless you strictly use the GPU for graphics accelerations, you must rewrite your code to run in the GPU. This is primarily because the CPU and the GPU use radically different programming models— SIMD versus multiple instruction streams, multiple data streams (MIMD). If you try to run code written for a MIMD processor like the CPU on a GPU, you are very likely to slow things down rather than speeding them up.

This rewrite or redesign of the code is a major drawback for the adoption of GPUs. We will see in the coming chapters that DL provides at least a partial answer to that challenge.

To give you an idea of the amount of rewriting that is required to move from the CPU to the GPU, we will go through a gentle introduction to the programming model used by the NVIDIA GPUs, followed by a couple of CUDA programs.

The programming model used by NVIDIA GPUs is called the Compute Unified Device Architecture (CUDA). The "unified" aspect applies only to NVIDIA. The model is unified across NVIDIA GPU devices, not across GPU devices from multiple vendors. Actually, the fallacy of the "unified" description is twofold: not only does it apply only to NVIDIA devices, but it also hides the fact that when you go from one generation of GPUs to the next, you often need to rewrite your code to take advantage of the performance enhancements provided by the hardware. That is clearly not optimal for a software developer.

As we discussed earlier, the big advantage of a GPU is the number of cores. To take advantage of the high number of cores, CUDA introduces the concept of a *kernel*. A kernel is simply a function that is executed by a CUDA core on a CUDA thread. Looking back at Figure 6.1, we see that the GPU is organized in multiple SMs that are further split into blocks. In CUDA, this hardware design is exposed in the form of blocks of threads. Each kernel runs in a thread, and each thread is in a block (of threads). As you might expect, because of the hardware design, the number of threads per block is limited to what the warp scheduler can handle. Today (in 2018), the limit on the number of threads is generally 1,024 per block. In other words, one can queue 1,024 threads for execution, but only 32 can run at the same time under the orchestration of the warp scheduler.

This hierarchy of blocks, threads, and kernels forces software developers to partition their problems into a number of threads per block and a number of blocks. For example, if we wanted to execute 10,000 draws to run Buffon's simulation, we would write something like the following code:

Program 6.1: Running Buffon's Simulation with 10,000 Draws

```
int nbr_of_threads = 10000;
int nbr_of_threads_per_block = 1024;
int nbr_of_blocks = nbr_of_threads / nbr_of_threads_per_block + 1;

buffon_kernel<<<nbr_of_blocks, nbr_of_threads_per_block>>> (...);
```

The `<<<nbr_of_blocks, nbr_of_threads_per_block>>>` code is a CUDA extension to the C language to indicate how many blocks of threads and how many threads (in each block) should be scheduled for execution. After the execution of this code, the body of the function (kernel), `buffon_kernel()`, is executed 10,240 times (notice that it is not exactly 10,000; more on this later).

With a GPU such as the Volta and its 5,120 CUDA cores, this means that we will have two consecutive parallel executions of 5,120 versions of `buffon_kernel()`. In each version, the code is the same, but the data is different. Obviously, to get over 5,000 threads, all SMs must participate. We will further examine this example shortly, but first, we must say a few words about the memory access in GPU coding.

An important point to understand (and to keep in mind) about the GPU is that the GPU is a device. In other words, you connect the GPU to your computer just as you would connect a hard-disk drive or a Blu-ray Disc player. The consequence is that a GPU is an addition to your computer, not the main show. This means that the decisions are still made by the CPUs running an operating system (OS) such as Windows or Linux. Practically, every time you want the GPU to do something on your behalf (for example, run thousands of threads), you must run an OS program that sends the data and the requests to the GPU card. The main contribution of CUDA is to greatly ease the communication between the OS program and the GPU card.

In your CUDA programs, this translates into the following division of labor:

- The OS or *host* part of your program runs like any program. As far the OS is concerned, there is nothing special about your CUDA program.
- The kernel or *device* part of your programs runs in the CUDA device. The OS doesn't "see" or do anything with your kernel. The CUDA run time transparently sends the data and the code for your kernel to the GPU device.

This difference between host and device has some important consequences regarding memory access in a CUDA program. The memory comes in multiple types, and some of those types are in short supply.

The different types of memory that you deal with as a CUDA programmer are organized in a hierarchy. There are four levels in that memory hierarchy, sorted here from the fastest to the slowest with respect to the GPU:

1. Local memory
   This memory is per thread. Each thread uses some of the memory for its local variables. Local memory is the fastest memory, but it is also the one in the shortest supply. The exact amount depends on the compute capability of your GPU card (the compute capability of a GPU defines its features). Typically, you have a few kilobytes (KBs = 10^3 bytes). That's right, these are KBs, not even megabytes (MBs = 10^6 bytes).

2. Shared memory
   This memory is per block of threads. It is similar to local memory in speed and availability (that is, you don't have much of it). Typically, you have tens of KBs.

3. Global memory
   This memory is for the entire card. It is global to all the thread blocks and is usually measured in gigabytes (GBs = 10^9 Bytes), at least on modern GPUs like the K80 that we use in this book.

4. Host memory
   This is the RAM available on your computer. It is typically measured in tens or hundreds of GBs, sometimes terabytes (TBs = 10^{12} Bytes).

The point for a CUDA developer is that memory is at a premium, both in its availability and in its speed. This implies that a good problem to handle on a GPU is a problem that requires a lot more (parallel) computations than memory. The exact quantification will be on a case by case basis and depends greatly on the skills of the CUDA programmer, but one can use common sense to have a rough idea. For example, a matrix multiplication is a very good candidate for GPU processing, because each matrix element requires several operations. By the same token, a matrix transposition is not a very good candidate for GPU processing, because each matrix element requires zero operations; you are only moving memory cells.

This concludes our very high-level and greatly simplified description of CUDA, the NVIDIA GPU programming model. When looking for optimal or near-optimal performance, other considerations must come into play. For example, how the memory is accessed plays a crucial role in performance. We won't go into those details in this book, since our goal is only to introduce GPUs and the CUDA programming model. Also, the problems that we tackle require little to no synchronization between the threads. An example of a problem that requires careful synchronization between threads is sorting. We didn't cover enough detail about CUDA to write a fast GPU-based sort.

For those interested in more details about CUDA and more complicated problems, they are covered in the *CUDA C Programming Guide* (NVIDIA 2017a).

Hello π

We will now go through a couple of straightforward examples of CUDA programming. Both problems have already been addressed in Chapter 5, so we can focus on the differences between SAS and CUDA programming.

The CUDA Toolkit

The first step in your CUDA development is to download the free CUDA Toolkit from https://developer.nvidia.com/cuda-downloads. There are versions for Windows, Mac OS X, and Linux.

We won't go over the steps to download and install the toolkit in this book, since the explanations from NVIDIA are straightforward and are more likely to stay up-to-date than the static content of a book.

For your development, we would recommend using NSight, an integrated development environment (IDE) that simplifies CUDA development over a command line-based interface. NSight is included with the CUDA toolkit.

Figure 6.2 shows a screen shot of NSight running in X Windows on OS X. The Mac and Linux versions are based on the Eclipse IDE, and the Windows version is based on the Visual Studio IDE.

Figure 6.2: NSight

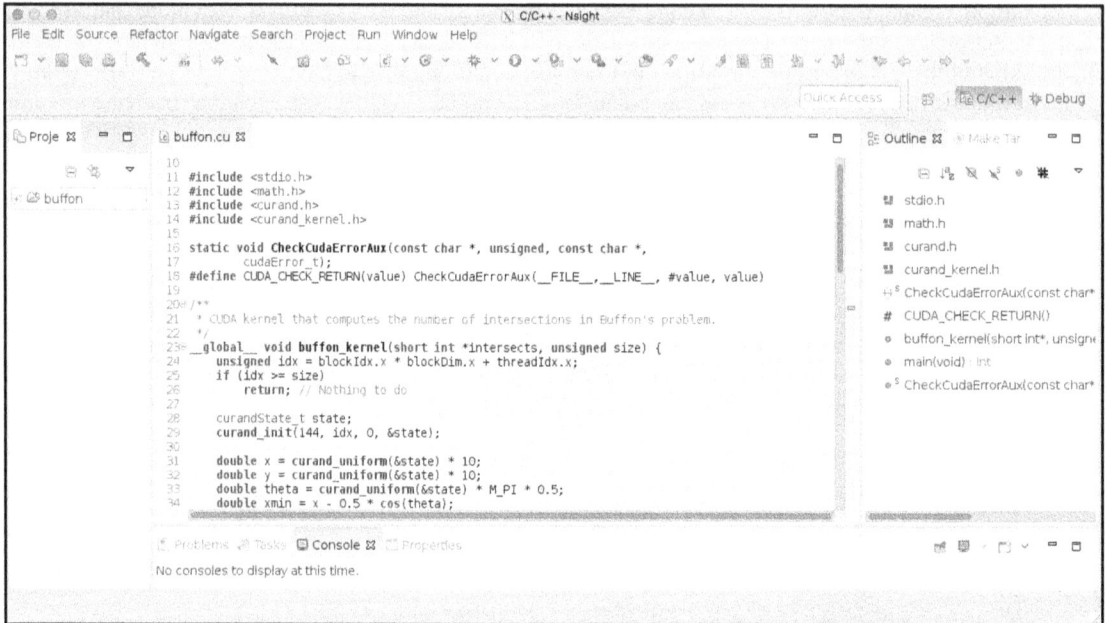

You might be surprised to learn that you don't need to have a CUDA device to install the CUDA toolkit. However, you need a CUDA device to run CUDA programs (there is no CUDA emulator). If you don't have a CUDA device on your personal machine, NSight supports remote development. For example, you can run NSight on a Mac and run your CUDA programs on a Linux machine.

Buffon Revisited

When we talked about Monte Carlo simulations in Chapter 5, we mentioned Buffon's needles and the potential use of this example to approximate the value of π. Let's revisit this simple example, but this time we will run it in thousands of threads on a GPU.

The code for this sample can be found in buffon.cu in the online code for Chapter 6. The cu extension is standard for CUDA programs. A CUDA program is in fact a C++ program, not a C program, but we mostly stick to C99 syntax in this chapter.

Like any good C program, we start by including a few header files:

Program 6.2: Including Header Files

```
#include <stdio.h>
#include <math.h>
#include <curand.h>
#include <curand_kernel.h>
```

The <stdio.h> and <math.h> are standard C header files for IOs and math operations (we will print the results and use a couple of math functions and constants). The next two files, <curand.h> and <curand_kernel.h> require a bit of explanation. We need to generate (pseudo-) random numbers for our simulations, but since we need to run the code in thousands of threads, the generation needs to be thread-safe and efficiently produce unique random numbers in all the threads. The standard C functions were not developed with these constraints in mind, so NVIDIA provides us with their versions. The usage is similar to the standard C functions:

- one function to initialize the random number generator. We use the `curand_init()` function.
- several functions to get random numbers. We use `curand_uniform()`, which returns a random number from a uniform distribution between 0 and 1.

We then define a function and a macro to check for CUDA errors, because many of the CUDA functions return a number (defined as a `typedef cudaError_t`):

Program 6.3: Defining a Function to Check for CUDA Errors

```
static void CheckCudaErrorAux(const char *, unsigned, const char *,
          cudaError_t);
#define CUDA_CHECK_RETURN(value) CheckCudaErrorAux(__FILE__,__LINE__, #value,
value)
```

The usage will become clear in a second.

We then define the kernel function with two arguments: the number of times a needle intersects a board and the number of needles. The kernel deposits the values in the memory cell that `intersects` points to):

```
__global__ void buffon_kernel(short int *intersects, unsigned size)
```

The keyword `__global__` is a CUDA extension to C. Its purpose is to state that the function must be compiled to generate code that can run on the CUDA device (remember, the kernel function is not executed by your CPU, so we don't want to generate code that could run on a Xeon, for example). The keyword `__global__` means that the kernel can be called by the CPU and by the GPU (more on this later).

The very first thing that we need to do in our kernel is check the thread identifier to make sure that the thread has work to do:

```
unsigned idx = blockIdx.x * blockDim.x + threadIdx.x;
if (idx >= size)
    return; // Nothing to do
```

The definition of `idx` and the test that follows requires a bit of explanation. As we mentioned previously, the warp scheduler can only schedule a fixed number of threads (typically 1,024). So if the number of threads that you need is not a multiple of 1,024, then you will have too many threads, and some threads will be left idle. It might seem strange to have too many threads and incur the cost of creating them, but on a GPU, creating a thread is extremely fast and cheap (in resources).

Now, let's look at the surprising use of `.x`, as in `blockIdx.x` and `blockDim.x`. The identifier for a block of threads and for a thread can have up to three dimensions, x, y, and z. (Here again the graphics origin of the GPU is apparent.) As in many GPGPU problems, we only use the first dimension in our example, so `blockIdx.x` is the identifier of the block assigned by CUDA. In practice, this block identifier is 0, 1, 2, 3, ... The same thing is true for `threadIdx.x`, which is the identifier inside the block (0, 1, 2, 3, ...). So by doing the preceding multiplication and addition, we obtain a globally unique thread identifier (across all threads of all blocks) that starts at 0 and stops at the number of threads minus 1. In our case, we test this global thread identifier against the number of needles (since we don't need to compute the intersection with the boards if the needle doesn't exist).

Next, we leverage the CUDA pseudo-random generator and generate the coordinates of the center of the needle (please refer to Chapter 5 for an explanation of the algorithm):

Program 6.4: Generating the Coordinates of the Center of the Needle

```
curandState_t state;
curand_init(clock(), idx, 0, &state);

double x = curand_uniform(&state) * 10;
double y = curand_uniform(&state) * 10;
double theta = curand_uniform(&state) * M_PI * 0.5;
```

It might seem ironic to use M_PI in this example, but our purpose is to show an example of CUDA code, not truly to come up with a good estimate of π. In the call to curand_init(), clock(), idx, and 0 play the role of a thread-aware seed (0 is a potential offset in the pseudo-random series, but we don't need it here).

As we did in the previous chapter in SAS code, we can now easily calculate whether the needle intersects a board or not:

Program 6.5: Calculating Whether a Needle Intersects a Board

```
   double xmin = x - 0.5 * cos(theta);
   double xmax = x + 0.5 * cos(theta);

   for (int lndx = 1; lndx < 10; lndx += 2) {
     if ((xmin < lndx) and (xmax > lndx)) {
       intersects[idx]++;
       break;
     }
   }
```

This is it for the kernel. The rest of the code that we need to write exclusively runs in the CPU.

Let's start with the main function and declare an array to accumulate the intersections of our needles:

```
int main(void) {
   const int size = 10000;
   short int *intersects = NULL;
```

We then allocate the memory for the needles:

```
   CUDA_CHECK_RETURN(cudaMallocManaged(&intersects,
                         sizeof(short int) * size));
```

Notice the use of the cudaMallocManaged() function, which has the advantage of allocating memory in the CPU (host) memory and in the GPU (global) memory. The word "global" is misleading (like a lot of the terminology in CUDA, starting with the acronym CUDA itself, as we indicated earlier). As we mentioned before, it is the memory on the GPU card, so it is global to all the CUDA threads, but not global to the computer that you're running on.

Prior to CUDA 6, when we didn't have managed memory, you had to allocate the host memory and the device (GPU) memory, and then you had to manually copy the data back and forth between the host and the device memory. Using memory managed by CUDA is a lot more convenient (and potentially faster, since the allocations and data movements are strictly on demand).

Now that we have memory for the CPU and the GPU, we calculate the number of blocks based on a fixed number of threads per block, 1,024:

```
int nbr_of_threads = size;
int nbr_of_threads_per_block = 1024;
int nbr_of_blocks = nbr_of_threads / nbr_of_threads_per_block + 1;
```

We can afford to use the maximum number of threads (1,024) because we don't need many resources per thread. If we had more demand on the local memory (with more code or automatic variables, for example) we would have to limit the number of threads to something less than 1,024. As you might expect, having fewer threads running at the same time implies less parallelism and worse performance. In this case, we can take full advantage of the parallelism of the GPU, as we will see shortly.

Don't forget to add one thread in case your number of threads is not a multiple of 1024!

We now have all the information we need to call the kernel:

```
buffon_kernel<<<nbr_of_blocks,
            nbr_of_threads_per_block>>>(intersects, size);

CUDA_CHECK_RETURN(cudaDeviceSynchronize());
```

The call to `cudaDeviceSynchronize()` is important: without it, we would not wait for the GPU to be done with its processing and our estimate of π would be ∞ most of the time.

The rest of the code is straightforward:

```
int total_intersects = 0;
for(int ndx = 0; ndx < size; ndx++) {
        total_intersects += intersects[ndx];
}

printf("PI estimate: %lf\n",
        (double)size / total_intersects);
```

Note that a more complicated implementation could perform the loop in the GPU, but that would require some synchronization between the different threads. For a loop with one addition, it is not worth it. Finally, don't forget to free the memory:

```
CUDA_CHECK_RETURN(cudaFree(intersects));
intersects = NULL;
```

Here again, CUDA manages the memory on the host and on the device.

If you let the program run with 10,000 draws or threads, you should see an output like the following in a fraction of a second:

```
PI estimate: 3.174603
```

If you bump up the number of needles to 10 million (each thread handles one needle), the execution does not take much longer and usually gives you a better estimate:

```
PI estimate: 3.140287
```

While the program is running, you can check what the device on your machine is doing. For example, you might see something like this if your machine has multiple devices:

```
[dl4na@fsnlax05 ~]$ nvidia-smi
Sun Jan 14 09:28:16 2018
+-----------------------------------------------------------------------------+
| NVIDIA-SMI 375.26                 Driver Version: 375.26                     |
|-------------------------------+----------------------+----------------------+
| GPU  Name        Persistence-M| Bus-Id        Disp.A | Volatile Uncorr. ECC |
| Fan  Temp  Perf  Pwr:Usage/Cap|         Memory-Usage | GPU-Util  Compute M. |
|===============================+======================+======================|
|   0  Tesla K80            Off | 0000:05:00.0     Off |                    0 |
| N/A   62C    P0   149W / 149W |    345MiB / 11439MiB |    100%      Default |
+-------------------------------+----------------------+----------------------+
|   1  Tesla K80            Off | 0000:06:00.0     Off |                    0 |
| N/A   38C    P0    73W / 149W |    108MiB / 11439MiB |      0%      Default |
+-------------------------------+----------------------+----------------------+

+-----------------------------------------------------------------------------+
| Processes:                                                       GPU Memory |
|  GPU       PID   Type   Process name                             Usage      |
|=============================================================================|
|    0    124096      C   .../dl4na/cuda-workspace/buffon/Debug/buffon  214MiB |
+-----------------------------------------------------------------------------+
```

As you can see, we're using 100% of one device. That was relatively easy to achieve because our problem is embarrassingly parallel.

To use both devices and run almost twice as fast, CUDA provides a API:

```
cudaError_t cudaSetDevice(int device)
```

In that function, the device number starts at 0. One strategy to use both devices could be to launch two kernels after setting the device and run the simulation on all cylinders. We will use that strategy shortly when we generate training data for the neural network that approximates the distance from the origin.

Note that CUDA provides absolutely no mechanism to resolve contention between multiple programs that attempt to set the device: you are on your own to provide some contention resolution. This lack of multi-user support shows the origin of the CUDA device: a single-user, single-purpose graphics accelerator card.

Now that we have seen a little of the mechanics of CUDA, let's use that experience for our real purpose: generating data for training a deep neural network.

Generating Random Walk Data with CUDA

The code for this example is available in the file `rw_training_data.cu`.

After the usual includes and error function definitions that we saw earlier in the Buffon sample, we define a thread-specific data structure that will hold the result of the random walk simulation:

Program 6.6: Defining the Data Structure

```
// Each GPU thread deals with the following data structure:
// num_steps, distance. All threads compute the same number
// of trials.
typedef struct thread_data_s {
  int num_steps;  // the number of steps in the random walk (INPUT)
  double distance;// the average distance for all trials.   (OUTPUT)
} thread_data;
```

In the output file, we also add the theoretical limit that we discussed in the previous chapter as a sanity check. We then enter the parameter of the random walks that we will simulate: the number of draws or trials for each random walk and the number of threads that we will schedule.

Program 6.7: Specifying the Number of Trials and Threads

```
static const int NUM_TRIALS  = 500; // For each random walk length,
                                     // we will try NUM_TRIALS times
static const int NUM_THREADS = 5000;// How many threads we fire
                                     // at the same time
```

We then define a helper function to compute a pseudo-random number between two values. This helper function soon comes in handy when we calculate the direction of the step that we take at each draw. You'll notice that we use the `__device__` CUDA extension to indicate that the function is available on the CUDA device only. There is no need to generate the code to run on the CPU. Only the GPU is needed, and the function can be called only from the GPU. Previously, we used the `__global__` CUDA extension to indicate that the function could run on the GPU, but be called from the CPU as well as from the GPU.

```
__device__ inline float rand_between(curandState_t state,
                         double min, double max) {
  return (min + floor((1 + max - min) * curand_uniform(&state)));
}
```

Next, we write the kernel. The beginning is similar to what we did before. We compute the (global) thread identifier and make sure that there is work to do in the current thread:

```
__global__ void kernel(int num_trials, thread_data* data) {
  unsigned idx = blockIdx.x * blockDim.x + threadIdx.x;
  if (idx < NUM_THREADS) {
```

If you recall the work that we did in the previous chapter using SAS, the next few lines are straightforward. We perform the random walk for the number of steps passed in the thread_data structure and return the distance from the origin in the same structure:

Program 6.8: Performing the Random Walk

```
                double distance_for_all_trials = 0;
                curandState_t state;
                curand_init(clock(), idx, 0, &state);
                for (int ndx_trials = 0; ndx_trials < num_trials;
                     ndx_trials++) {
                   double x = 0;
                   for (int ndx_step = 0; ndx_step < data[idx].num_steps;
                        ndx_step++) { /* We walk data[idx].num_steps */
                      double x_step =
                         curand_uniform(&state) <= 0.5 ? -1 : 1;
                      x += x_step;
                   }
                   distance_for_all_trials += sqrt(x * x);
                }

                data[idx].distance = distance_for_all_trials / num_trials;
```

Note that to get different sequences every time, we use the clock() function as a seed. We want a different sequence every time in this case, since the goal is to get more training data. You typically don't want a different sequence if you're debugging. In that case, you would use a constant as the seed.

This completes the code that needs to run on the GPU, so we are ready to write the main function, which runs on the CPU under the supervision of the OS like any of your SAS programs.

We define a few arguments so that we can generate the training data on multiple devices. If you run the executable without any arguments (or the wrong arguments), you get the Help (slightly reformatted for clarity):

```
[dl4na@fsnlax05 data]$ ./rw_training_data
Invalid number of arguments: 1. Expecting 4.
nw_training_data: generate training data for learning the distance from
               the origin in a random walk.
Usage: nw_training_data <gpu-device-id> <min-num-steps> <max-num-steps>
      gpu-device-id = 0|1 depending on which device to use.
      min-num-steps = any positive value.
      max-num-steps = any positive value bigger than min-num-steps.
We always compute and report a multiple of 5000 values.
```

We will skip reviewing the pure C code that prints the preceding usage and parses the arguments, since it doesn't have anything to do with CUDA development. Things get more interesting in the main function once we have the input arguments that we expect. First, we set the device using the API that we discussed earlier. This enables us to easily run multiple instances in parallel, one for each GPU device that we have on our machine:

```
      CUDA_CHECK_RETURN(cudaSetDevice(device_id));
```

We then allocate the memory for the thread_data structure. Notice that here again, we rely on memory managed by CUDA:

```
thread_data* data = NULL;
CUDA_CHECK_RETURN(cudaMallocManaged(&data,
                   NUM_THREADS * sizeof(thread_data)));
```

To generate the data, we will loop over the number of steps that we want to simulate. Once again, the goal is to have one record per random walk simulation of a given number of steps. Each record that we produce will have information like the following:

Number of Steps	Distance from the Origin
499991	567.9520

Consequently, we need one loop for each record and then one nested loop for each thread (or draw). Specifically, the code in the next three boxes loops through all records to calculate their distance to the origin, starting with the initialization of the data structures used by the GPU threads:

```
for(int num_steps = min_num_steps; num_steps < max_num_steps;) {
    for (int ndx = 0; ndx < NUM_THREADS; ndx++) {
        data[ndx].num_steps = num_steps++;
        data[ndx].distance = -1;
    }
```

Note that the records are computed in batches of NUM_THREADS to maximize the usage of GPU threads. There is also a bit of code to open the output file that we don't review here.

At this point, we know how many threads we want to run, so let's run them (don't forget to wait for the GPU to be done with all the threads):

```
static const int BLOCK_SIZE = 1024;
const int blockCount = max(1, NUM_THREADS / BLOCK_SIZE + 1);

kernel<<<blockCount, BLOCK_SIZE>>>(NUM_TRIALS, data);
CUDA_CHECK_RETURN(cudaDeviceSynchronize());
```

Except for writing the results and freeing the memory, we are essentially done. Please note that there are in fact hidden copies of data from the CPU (main) memory to the GPU and vice versa, but those are transparent to you thanks to CUDA managing the memory:

```
for(int ndx = 0; ndx < NUM_THREADS; ndx++) {
    if(-1 != data[ndx].distance) {
        double limit = sqrt(2 * data[ndx].num_steps / M_PI);
        fprintf(training_data_file,
            "%7d, %10.4lf, %10.4lf\n", data[ndx].num_steps,
            data[ndx].distance, limit);
    }
}
fflush(training_data_file);
}

CUDA_CHECK_RETURN(cudaFree(data));
data = NULL;
```

As mentioned previously, we also write the theoretical limit of the distance from the origin.

Running the code simply involves passing in a few arguments:

- the GPU device ID, starting at 0.
- the minimum number of steps for the random walk.
- the maximum number of steps for the random walk. We will run one draw for each number of steps to be considered.

Notice that we can run multiple instances in parallel, one per device that we have on our machine:

```
[dl4na@fsnlax05 data]$ ./rw_training_data  0 300000 500000 &
[1] 31712
[dl4na@fsnlax05 data]$ ./rw_training_data  1 300000 500000 &
[2] 31778
```

After starting the instances of rw_training_data, you should start seeing the printouts for each instance (the first set is for device 0, and the second set is for device 1):

```
rw_training_data: Starting...
 min_num_steps = 300000, max_num_steps = 500000, NUM_TRIALS = 500
 Initializing the thread data...
 Calling the kernel for 300000 to 304999 steps with 5 blocks of 1024 threads at
15:28:11

rw_training_data: Starting...
 min_num_steps = 300000, max_num_steps = 500000, NUM_TRIALS = 500
 Initializing the thread data...
 Calling the kernel for 300000 to 304999 steps with 5 blocks of 1024 threads at
15:29:12
```

If you look at the running processes, you should see two instances of rw_training_data (shown here for the Linux operating system):

```
[dl4na@fsnlax05 data]$ ps
  PID TTY          TIME CMD
30925 pts/6    00:00:00 bash
31712 pts/6    00:05:08 rw_training_dat
31778 pts/6    00:04:06 rw_training_dat
32051 pts/6    00:00:00 ps
```

If you check the state of the GPUs, you should see each of them running at 100%:

```
[dl4na@fsnlax05 data]$ nvidia-smi
Sun Jan 21 15:29:32 2017
+-----------------------------------------------------------------------------+
| NVIDIA-SMI 375.26                 Driver Version: 375.26                     |
|-------------------------------+----------------------+----------------------+
| GPU  Name        Persistence-M| Bus-Id        Disp.A | Volatile Uncorr. ECC |
| Fan  Temp  Perf  Pwr:Usage/Cap|         Memory-Usage | GPU-Util  Compute M. |
|===============================+======================+======================|
|   0  Tesla K80           Off  | 0000:05:00.0     Off |                    0 |
| N/A   52C    P0   104W / 149W |     66MiB / 11439MiB |    100%      Default |
+-------------------------------+----------------------+----------------------+
|   1  Tesla K80           Off  | 0000:06:00.0     Off |                    0 |
| N/A   40C    P0   116W / 149W |     66MiB / 11439MiB |    100%      Default |
+-------------------------------+----------------------+----------------------+

+-----------------------------------------------------------------------------+
| Processes:                                                       GPU Memory |
|  GPU       PID   Type   Process name                             Usage      |
|=============================================================================|
|    0     31712      C   ./rw_training_data                            62MiB |
|    1     31778      C   ./rw_training_data                            62MiB |
+-----------------------------------------------------------------------------+
```

This will run for a while (depending on your hardware), and at the end of the execution, you should see that the results of the simulations have been written to disk. You should see the following for the first instance:

```
data has been written to training_data_0.csv in folder /data
rw_training_data: All done!
   Done at 15:56:24.
```

And you should see the following for the second instance:

```
data has been written to training_data_1.csv in folder /data
rw_training_data: All done!
```

For each GPU, we can run 5,000 draws (around 300,000 steps) in about 30 seconds (fewer steps will be faster, more steps will be slower):

```
Calling the kernel for 300000 to 304999 steps with 5 blocks of 1024 threads at
11:31:20
   Done at 11:31:52.
```

You should also see the two files:

```
[dl4na@fsnlax05 data]$ ls -ls *.csv;wc -l *.csv; head *.csv
6284 -rw-r--r-- 1 dl4na sas 6400000 Jan 21 15:56 training_data_0.csv
6284 -rw-r--r-- 1 dl4na sas 6400000 Jan 21 15:57 training_data_1.csv


  200000 training_data_0.csv
  200000 training_data_1.csv
  400000 total

==> training_data_0.csv <==
300000,    408.9440,    437.0194
300001,    419.2680,    437.0201
300002,    442.8600,    437.0208
300003,    412.8800,    437.0216
300004,    455.5080,    437.0223
300005,    445.9640,    437.0230
300006,    433.2920,    437.0237
300007,    441.3560,    437.0245
300008,    442.0320,    437.0252
300009,    450.5880,    437.0259

==> training_data_1.csv <==
300000,    439.1520,    437.0194
300001,    421.2880,    437.0201
300002,    443.5080,    437.0208
300003,    451.8280,    437.0216
300004,    439.9400,    437.0223
300005,    443.7120,    437.0230
300006,    422.7800,    437.0237
300007,    425.9360,    437.0245
300008,    422.4400,    437.0252
300009,    446.0800,    437.0259
```

Putting It All Together

Now that we have developed a couple of CUDA programs, let's see how we can call them from SAS. In this chapter, we call CUDA code from a single-threaded SAS program. In the next chapter, we will marry many-task computing (MTC) and CUDA for the fastest possible analytics.

To call the CUDA code from SAS, we must define a shared object (.SO file on Linux) or a dynamically linked library (.DLL on Windows). We've included an NSight project that builds a shared object for Linux as an example. In addition to the shared object, we must have an entry point that we can declare in SAS using the PROTO procedure (more on this later). This requires a minor refactoring of the code in rw_training_data.cu.

First, we declare the prototype of a new function, rw_training_data. We must declare this as a C function, because CUDA files are in fact C++ files, and we don't want the C++ declaration of the names:

Program 6.9: Declaring the rw_training_data Function

```
extern "C" {
  void rw_training_data(int device_id, /* IN The id of the device*/
          int min_num_steps,          /* IN Min number of steps */
          int max_num_steps);         /* IN Max number of steps */
};
```

We can then call `rw_training_data` from main:

Program 6.10: Calling the Function

```
int main(int argc, char *argv[]) {
...
    printf("rw_training_data: Starting...\n");
    rw_training_data(device_id, min_num_steps, max_num_steps);
    printf("rw_training_data: All done!\n");

    return 0;
}
```

The body of `rw_training_data` is the exact same code.

After you build the NSight project (or manually build the shared object with the included makefile), the resulting shared object is put in `/local/lib` for easy reference (short paths are more practical).

Armed with a shared object (or a dynamically linked library), we can write a bit of SAS code to load and run the GPU code from SAS. The code is in `dl_run_rw_training_data.sas`, which you can find in the code folder of Chapter 6.

Let's briefly review the SAS code. Note that there is absolutely nothing specific to the GPU in this code. The SAS software has no knowledge that under the `rw_training_data` function there is really a GPU kernel running in thousands of threads.

We pass the arguments that we need on the SAS command line (this is similar to what we did with the previous C program):

```
[dl4na@fsnlax05 ~]$ /data/local/install/SASHome/SASFoundation/9.4/sas -stdio
dl_run_rw_training_data.sas -set device_id 0 -set min_num_steps 300000 -set
max_num_steps 500000
```

The SAS program then starts by assigning a library name to the location of the shared object (we removed most of the log for clarity):

```
...
Calling the generation of training on device 0 in [300000, 500000]
11
12          libname lib "/local/lib/";
NOTE: Libref LIB was successfully assigned as follows:
      Engine:      V9
Physical Name: /local/lib
```

Now that we have defined the library, we can load the shared object:

```
14          proc proto package=lib.proto_ds.rw_data
15                  label="Random Walk Training Data Generators";
16
17          link "/local/lib/librw_training_data_lib.so";
18
19          void rw_training_data(int device_id,
20                  int min_num_steps,
21                  int max_num_steps);
22      run;

NOTE: '/local/lib/librw_training_data_lib.so' loaded from specified path.
NOTE: Prototypes saved to LIB.PROTO_DS.RW_DATA.
NOTE: PROCEDURE PROTO used (Total process time):
      real time           0.00 seconds
      cpu time            0.00 seconds
```

Next, we define an FCMP function that calls the C function, because a DATA step can call only FCMP subroutines:

```
24          proc fcmp inlib =lib.proto_ds
25                  outlib=lib.fcmp_ds.rw_data;
26
27          subroutine sas_rw_training_data(device_id,
                              min_num_steps, max_num_steps);
28              call rw_training_data(device_id, min_num_steps,
                              max_num_steps);;
29          endsub;
30
31      quit;

NOTE: Function sas_rw_training_data saved to lib.fcmp_ds.rw_data.
NOTE: PROCEDURE FCMP used (Total process time):
      real time           0.02 seconds
      cpu time            0.03 seconds
```

Now that we have an FCMP subroutine that calls the C function that launches the CUDA kernels, we can start the generation of the data (notice that the output of the C function that launches the kernel appears in the SAS log):

```
35          data _null_;
36          call sas_rw_training_data(%sysget(device_id),
                          %sysget(min_num_steps), %sysget(max_num_steps));
37      run;
 min_num_steps = 300000, max_num_steps = 500000, NUM_TRIALS = 500
 Initializing the thread data...
 Calling the kernel for 300000 to 304999 steps with 5 blocks of 1024 threads at 17:25:05
   Done at 17:25:37.
 ...
 Initializing the thread data...
 Calling the kernel for 495000 to 499999 steps with 5 blocks of 1024 threads at 17:51:45
   Done at 17:52:36.
   ----
   data has been written to training_data_0.csv in folder /home/dl4na

NOTE: DATA statement used (Total process time):
      real time          27:31.89
      cpu time           27:32.19
```

For another example of SAS calling GPU code, please see the study by Bequet and Chen (2017). In that example, CUDA is not used to generate training data, but to very rapidly compute analytics for an insurance company.

Conclusion

We started this chapter by looking at the history of the GPUs. We saw that the path from the graphics card to the GPU to the GPGPU that we use in DL was not a straight one, and that the current GPGPU retains characteristics from the early graphics processing days.

Nevertheless, we saw that the GPU is one of the best SIMD processors that money can buy today. So we made use of it, first on a simulation of Buffon's problem and then on generating training data to train a neural network to approximate the distance from the origin of a random walk, given the number of steps.

We had a cursory look at CUDA, the programming model supported by NVIDIA GPUs. We experienced first-hand that CUDA development can be challenging at times, since it can be very close to the hardware.

In the next chapter, we will leverage the training data that we generated in this chapter to train a deep neural network.

Chapter 7: Monte Carlo Simulations with Deep Learning

In this chapter, we revisit the random walk example that we introduced in Chapter 5. Our goal is to significantly improve the performance of Monte Carlo simulations by using a trained artificial neural network (ANN) to approximate the results of the Monte Carlo simulations.

The example that we are trying to solve is straightforward: given a number of steps of a random walk, estimate the distance from the origin.

Generating Data

For the purpose of this book, we confine ourselves to random walks of 300,000 to 500,000 steps in one dimension. Extending the domain (longer random walks) or the number of dimensions (more random walks) would be straightforward and only require more computing power.

Training Data

To generate the training data, we rely on the CUDA program that we developed in the previous chapter. As discussed in that chapter, we run the program for each GPU device that we have, in our case two:

```
[dl4na ~]$ sas -stdio dl_run_rw_training_data.sas -set device_id 0 -set
min_num_steps 300000 -set max_num_steps 500000
...
[dl4na ~]$ sas -stdio dl_run_rw_training_data.sas -set device_id 1 -set
min_num_steps 300000 -set max_num_steps 500000
...
```

Each of those runs generates 200,000 observations. To have plenty of data for training and testing, we'll generate a total of 1,000,000 observations, which will be more than enough for training our deep neural network (DNN). As we discussed in Chapter 2, a DNN is an ANN with two or more hidden layers.

The 1 million observations is a bit of overkill, because, as we will see shortly, 100,000 observations are all we need for training the DNN in this case (the function to learn is simple). For convenience, we combine the two sets of training data into one:

```
[dl4na ~]$ echo "nb_steps, distance, limit" >training_data.csv
[dl4na ~]$ cat training_data_[01].csv | sort >>training_data.csv
[dl4na ~]$ wc -l training_data.csv
1000001 training_data.csv
```

And we can see that the overall file is what we expect, namely multiple observations for each number of steps:

```
[dl4na ~]$ head training_data.csv; tail training_data.csv
nb_steps, distance, limit
 300000,   420.7760,   437.0194
 300000,   432.6360,   437.0194
 300000,   434.9040,   437.0194
 300000,   442.0440,   437.0194
 300000,   446.0600,   437.0194
 300001,   409.1160,   437.0201
 300001,   432.6120,   437.0201
 300001,   433.5720,   437.0201
 300001,   451.6200,   437.0201
 499998,   525.8920,   564.1885
 499998,   546.0200,   564.1885
 499998,   559.8960,   564.1885
 499998,   574.3320,   564.1885
 499998,   577.6800,   564.1885
 499999,   540.6960,   564.1890
 499999,   549.5320,   564.1890
 499999,   557.9920,   564.1890
 499999,   560.7120,   564.1890
 499999,   602.4480,   564.1890
```

As we observed previously, there is some variability in the results returned by the Monte Carlo simulations. We will attempt to quantify that a bit in a few paragraphs.

The dl_load_rw.sas file is available in the code folder of this chapter. It contains a simple SAS program that examines the data and loads it into CAS so that we can later learn the weights and biases from that data.

Let's have a quick look at the code. It is similar to what we did before when looking at non-stochastic data.

First, we need a CAS session, since we rely on CAS to do our deep learning (DL) modeling:

Program 7.1: Initiating a CAS Session

```
options cashost='fsnlax05' casport=5570;
cas mysession;
libname CASLIB cas sessref = mysession;
```

CASLIB is our libref to CAS, but before loading the data into CAS, we import our CSV training data into WORK:

Program 7.2: Importing Training Data

```
proc import datafile='~/training_data.csv'
            out=function_training;
run;
```

We then augment the training data set with the difference between the distance that we estimated with the Monte Carlo simulations and the theoretical limit that we discussed in Chapter 5:

Program 7.3: Computing Differences

```
data CASLIB.function_training;
  set function_training;
  difference = abs(limit - distance);
run;
```

This additional field enables us to gauge the validity of the estimates produced by our DNN. Next, we compute and print the mean of that difference:

Program 7.4: Computing the Mean of the Difference

```
proc summary data=CASLIB.function_training nway;
  var difference;
  output out=function_training_summary (drop=_:) mean=;
run;

title 'mean(abs(limit - distance))';
proc print data=function_training_summary;
run;
title;
```

Now we go back to our regular pattern of loading the data into CAS and examining it:

Program 7.5: Loading Data into CAS and Examining It

```
proc cas;
  table.save/
    table = "function_training"
    name  = "function_training"
    replace = TRUE;
quit;

/* Have a look at the data we loaded into CAS */
proc casutil; list;
run; quit;

ods graphics / reset width=6.4in height=4.8in imagemap;

proc surveyselect data=CASLIB.function_training
   method=srs n=1000 out=function;
run;
```

```
proc sgplot data=function;
  title 'Random Walk for Deep Learning';
  YAXIS LABEL = 'Distance from the Origin'
        GRID VALUES = (350 TO 650 BY 10);
  scatter x=nb_steps y=distance;
run;

ods graphics / reset;
```

We rely on the SURVEYSELECT procedure (PROC SURVEYSELECT) to get a subset of the data to plot and to print. As you can see, we plot more than we print (1000 versus 20):

Program 7.6: Plotting and Printing Data

```
proc surveyselect data=CASLIB.function_training
   method=srs n=20 out=function;
run;
proc print data=function; run;

cas mysession terminate;
```

Now we're done with loading, plotting, and printing the data. So let's have a look at the outputs of our SAS program.

As Figure 7.1 shows, we've observed a mean difference of 13.57 between the Monte Carlo simulation results and the theoretical limits. We will keep this difference in mind when we run the inference through our DNN with our testing data.

Figure 7.1: Monte Carlo versus Theoretical

Obs	difference
1	13.5951

In Figure 7.2, we see the training data in CAS. There are no surprises here; we have a million records at our disposal.

Figure 7.2: Training Data in CAS

Caslib Information	
Library	CASUSER(dl4na)
Source Type	PATH
Description	Personal File System Caslib
Path	/home/dl4na/casuser/
Session local	No
Active	Yes
Personal	Yes
Hidden	No
Transient	Yes

Table Information for Caslib CASUSER(dl4na)							
Table Name	Number of Rows	Number of Columns	Indexed Columns	NLS encoding	Created	Last Modified	Promoted Table
FUNCTION_TRAINING	100000	4	0	utf-8	22Mar2018:12:06:49	22Mar2018:12:06:49	No

Table Information for Caslib CASUSER(dl4na)			
Table Name	Repeated Table	View	Compressed
FUNCTION_TRAINING	No	No	No

In Figure 7.3, we see the results of the plot of a sample of the data. Looking at the plot, you might think that the relationship between the number of steps and the distance from the origin is linear, but we know from Chapter 5 that the relationship is approximately $\sqrt{2 \cdot n / \pi}$ for large values of the number of steps (n).

Figure 7.3: Distance from the Origin

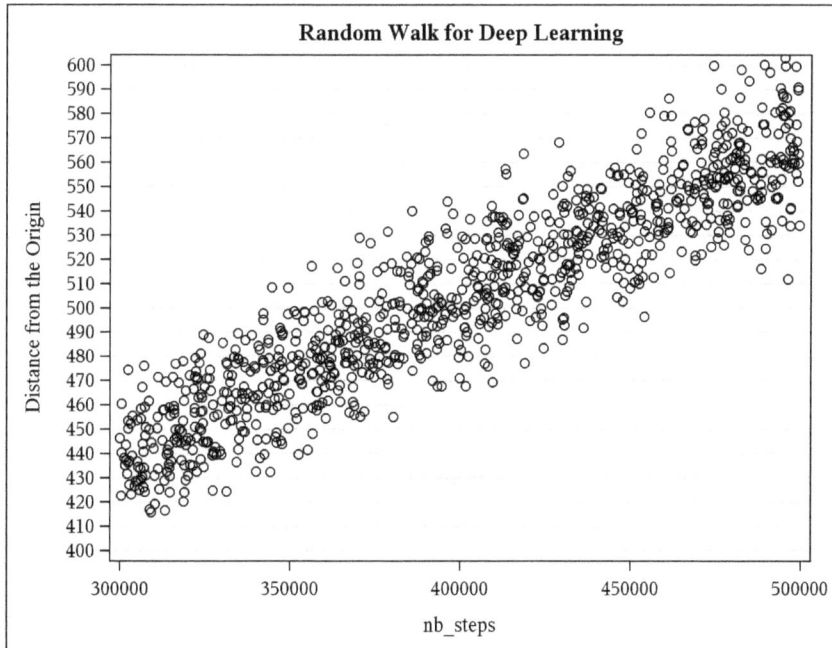

Finally, in Figure 7.4, we show the results of PROC PRINT on a sample of 20 observations. Here again, there are no surprises: the data is as we expected it.

Figure 7.4: Distance from the Origin with Difference

Obs	nb_steps	distance	limit	difference
1	307165	453.212	442.2073	11.0047
2	323191	443.796	453.5965	9.8005
3	324346	448.164	454.4063	6.2423
4	324515	464.44	454.5247	9.9153
5	330170	465.032	458.4678	6.5642
6	336331	461.788	462.7256	0.9376
7	339898	439.368	465.1729	25.8049
8	343860	463.96	467.8761	3.9161
9	399595	478.352	504.371	26.0190
10	401391	506.996	505.5032	1.4928
11	423250	535.908	519.0851	16.8229
12	429838	506.828	523.1093	16.2813
13	438100	511.888	528.1128	16.2248
14	440883	538.68	529.7875	8.8925
15	448766	537.152	534.5029	2.6491
16	448930	512.28	534.6005	22.3205
17	460263	530.532	541.3063	10.7743
18	463762	554.308	543.36	10.9480
19	470126	555.636	547.0754	8.5606
20	497487	592.816	562.77	30.0460

Now that we have training data, we need testing data.

Testing Data

To generate testing data, we simply generate a new set of the training data:

```
[dl4na ~]$ cd /tmp

[dl4na t/mp]$ echo "nb_steps,distance,limit" >training_data_1.csv

[dl4na /tmp]$ /data/local/install/SASHome/SASFoundation/9.4/sas
-stdio ~/dl_run_rw_training_data.sas -set device_id 1
-set min_num_steps 300000 -set max_num_steps 500000

[dl4na /tmp]$ mv training_data_1.csv ~/testing_data.csv
```

Notice the trick of adding a header line before the training data. It works because we append data when we generate the simulation results. In addition, please note that we could also have added an extra argument to our CUDA program to save us the file rename operation:

```
Let's check that the testing data is what we expect:
[dl4na /tmp]$ cd ~

[dl4na@fsnlax05 ~]$ wc -l testing_data.csv
200001 testing_data.csv

[dl4na@fsnlax05 ~]$ head testing_data.csv; tail testing_data.csv
nb_steps,distance,limit
 300000,    440.3400,    437.0194
 300001,    420.2680,    437.0201
 300002,    446.5520,    437.0208
 300003,    418.9440,    437.0216
 300004,    448.3600,    437.0223
 300005,    456.5120,    437.0230
 300006,    434.3680,    437.0237
 300007,    409.3160,    437.0245
 300008,    428.9440,    437.0252
 499990,    594.6320,    564.1839
 499991,    562.9280,    564.1845
 499992,    563.7160,    564.1851
 499993,    528.1720,    564.1856
 499994,    559.4280,    564.1862
 499995,    548.5640,    564.1868
 499996,    528.5560,    564.1873
 499997,    551.9800,    564.1879
 499998,    533.6480,    564.1885
 499999,    545.8560,    564.1890
```

This is what we expect—one observation per number of steps from 300,000 to 500,000 steps (with no repetitions) for a total of 200,000 observations.

Let's also do a cursory check to make sure that we don't have a complete overlap of the testing and training data by looking in the training data for a specific testing data distance, `559.4280`:

```
[dl4na@fsnlax05 ~]$ grep '559.4280,' testing_data.csv
 451446,    559.4280,    536.0965
 472589,    559.4280,    548.5066
 499994,    559.4280,    564.1862
[dl4na@fsnlax05 ~]$ grep '559.4280' training_data.csv
 436334,    559.4280,    527.0473
 452746,    559.4280,    536.8678
 452830,    559.4280,    536.9176
 458080,    559.4280,    540.0211
 460589,    559.4280,    541.4980
 463580,    559.4280,    543.2533
 468518,    559.4280,    546.1390
 468835,    559.4280,    546.3237
 472244,    559.4280,    548.3064
 473500,    559.4280,    549.0350
 476942,    559.4280,    551.0270
 480073,    559.4280,    552.8327
 482669,    559.4280,    554.3254
 482907,    559.4280,    554.4620
 485489,    559.4280,    555.9424
 485902,    559.4280,    556.1788
```

There are a few overlaps, but none for the same number of steps, so that's fine. A Monte Carlo simulation for different numbers of steps sometimes returns the same distance from the origin, so that is probable.

Everything looks good. We will score that data once we're done training.

Training the Network

The training code is very similar to what we did for the function regression in Chapter 3. Let's quickly review the new version of the code and point out the few differences.

As usual, we log in to CAS and define a caslib:

Program 7.7: Defining a Caslib

```
options cashost='fsnlax05' casport=5570;
cas mysession;

libname DL_LIB cas sessref = mysession;
```

We then launch PROC CAS and load the DL framework that we developed in Chapter 3:

Program 7.8: Loading the DL Framework

```
proc cas;
%include "~/model_io.sas";   /* load_model, etc. */
%include "~/ua.sas";         /* inference, train, etc. */

/* We load the data */
training_table   = "function_training.sashdat";
tables           = setup_training_data({name = training_table}, 80);
training_table   = tables.training;
validation_table = tables.validation;
```

The model name needs to change, but the rest stays the same:

Program 7.9: Specifying the Model Name

```
/* We build the model */
model_name              = "distance_from_origing_in_rw_model";
number_of_hidden_layers = 8;
define_model(model_name, number_of_hidden_layers);
```

With the model defined, we are now ready to train it for our new function:

$$distance = f(nb\_steps)$$

Program 7.10: Training the Model

```
inputs              = {"nb_steps"};
target              = "distance";
categorical_inputs  = { };
use_gpu             = 1;
max_epoch           = 100;
train(model_name, training_table, validation_table,
      inputs, target, categorical_inputs, max_epoch, use_gpu);
```

Once the training is done, we run the inference on the training set to plot it. As you can see, we want to preserve the inputs and outputs of the training table, including the limit, since we are interested in evaluating the difference between the theoretical limit and the distance from the origin as estimated by our DNN:

Program 7.11: Running the Inference on the Training Set

```
scoring_table = "function_scoring";
copy_vars = { "nb_steps", "distance", "limit" };
inference(model_name, training_table, scoring_table, copy_vars);
```

The rest of the code is like what we did in Chapter 3, except for the scoring plots that use a scatter plot (in **bold**):

Program 7.12: Plotting the Training Set

```
proc sgplot data=function_scoring;
  title "Truth versus Prediction (Training Set)";
  YAXIS LABEL = 'Distance from the Origin'
       GRID VALUES = (350 TO 650 BY 10);
  scatter x=nb_steps y=distance;
  scatter x=nb_steps y=_DL_Pred_;
run;
```

A scatter plot works better than a line plot, because we have multiple observations for a given number of steps in the training data. The titles and axes were also adjusted to reflect the new function that we are approximating in this case.

After a few epochs (100), things are starting to take shape, as you can see in Figure 7.5 and Figure 7.6.

Figure 7.5: Training after 100 Epochs

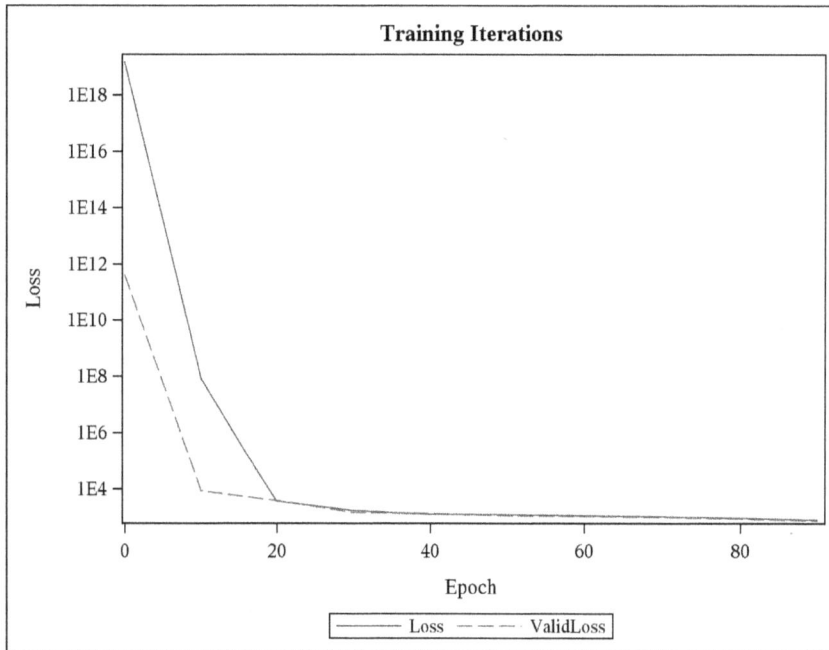

This is like what we observed in Chapter 3: at the beginning of the training, the DNN learns a linear regression.

Figure 7.6: Accuracy after 100 Epochs

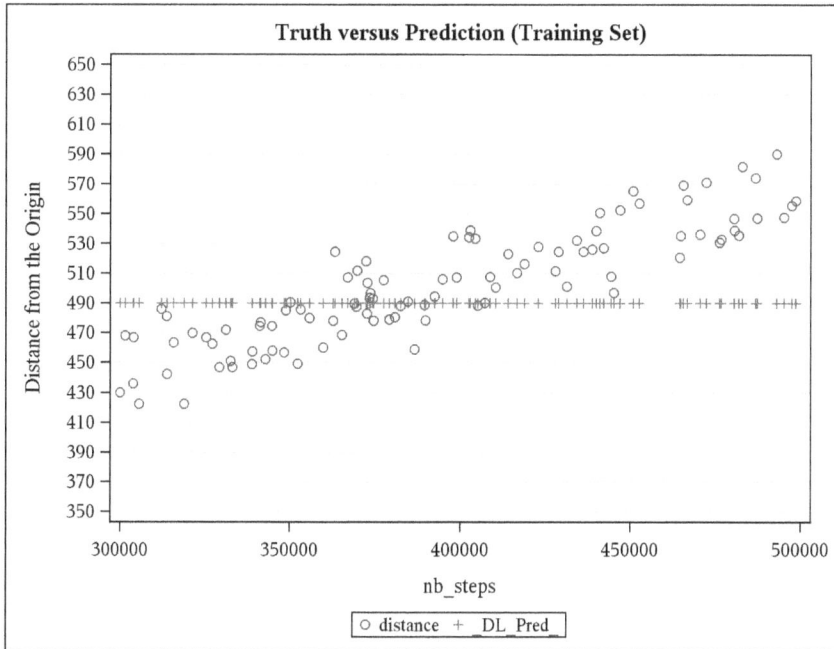

If you look at Figure 7.5, you'll notice that the loss tends to level off at around 60 or 80 epochs. That's somewhat concerning. As we discussed in Chapter 2, contrary to common wisdom, steep learning curves are your friend. The tail end is most definitely not steep, so that is not ideal. It would be closer to ideal if we had a better fit in Figure 7.6.

Another concerning fact about the curve in Figure 7.6 is its somewhat erratic nature. This is very different from what we saw in Chapter 3 when we ran a regression. (See Chapter 3, Figure 3.3.) Furthermore, in this case, we are trying to learn the approximation of a simpler function:

$$f(n) = \sqrt{2 \cdot n / \pi}$$

Compare this to the function in Chapter 3:

$$f(x) = \sin(x) * pdf('normal', x, 0, 2)$$

Perhaps things will get better as we train more. So, let's double the number of epochs to 200 to see whether we like those results better.

In Figure 7.7, you can see the loss after 200 epochs. This is not good at all; we are not learning anything. We are trapped in the (wrong) linear regression that we see in Figure 7.6. In other words, more epochs didn't help. We are stuck in a rut.

Figure 7.7: Training after 200 Epochs

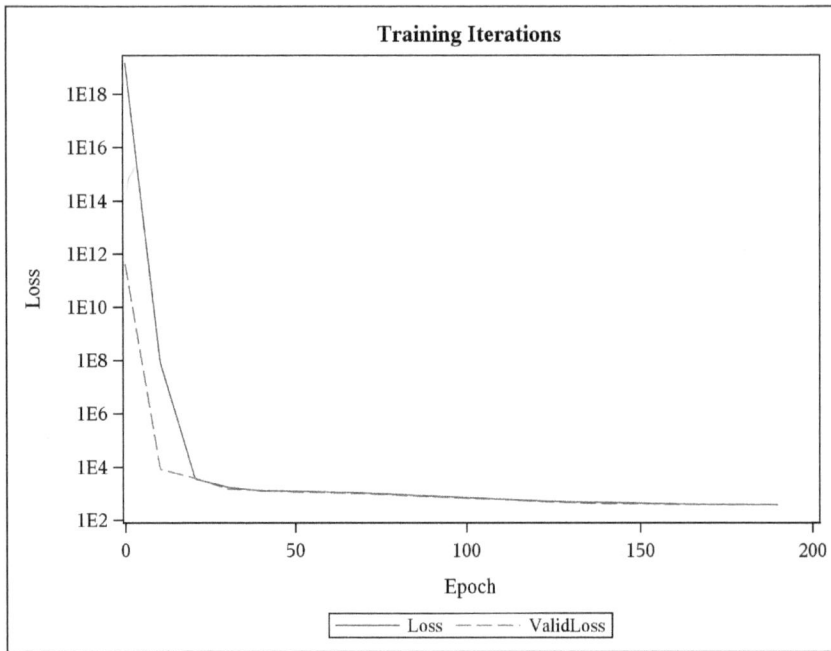

What could be the issue? We are learning a simpler function than in Chapter 3, so we ought to be able to replicate that learning in this case. Why are we having a tough time converging? Well, one thing that is clearly different between what we did in Chapter 3 and the function that we are trying to learn now is the range. In Chapter 3, all inputs and outputs were close to the origin: the feature values (*x*) were in [-10, +10] and the function values were in [-0.2, 0.2].

Let's normalize the inputs and outputs around the origin and see whether our learning abilities improve. The SAS Infrastructure for Risk Management developers conveniently implemented a macro to normalize the data, so we use that macro in dl_load_rw.sas:

Program 7.13: Normalizing the Data

```
%include "irm_wordcount.sas";
%include "irm_normalize_data.sas";

%irm_normalize_data(in_tbl       =full_function_training,
                    var_lst      =nb_steps distance limit,
                    out_tbl_stat =full_function_training_stat,
                    out_tbl      =full_function_training);
```

Note that the irm_normalize_data macro conveniently saves the mean and standard deviation of the normalized variables, as you can see in Table 7.1. These values will come handy when we run the inference through our trained DNN, so we save them in CASLIB(dl4na) under the name full_function_training_stat.

Table 7.1: Normalization Statistics

NAME	MEAN	STD
nb_steps	399999.5	57735.055786
distance	503.29240602	40.396735398
limit	503.29628237	36.612075613

Let's restart learning for 100 epochs and see how things are. After a few seconds, we get the training losses that you can see in Figure 7.8.

Figure 7.8: Normalized Training after 100 Epochs

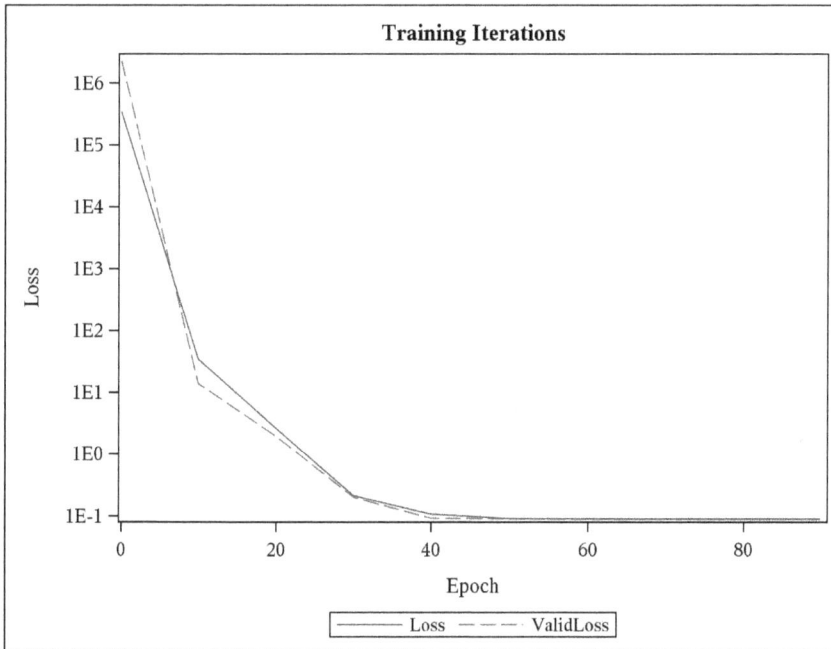

This is much better! In Figure 7.9, you see that the improved learning curve translates into a better model, even after 100 epochs. Not only is the accuracy better after only a few epochs, but you can start seeing the nonlinearity of a parabolic curve in Figure 7.9 as well. We could also have stopped early in the training process; at 50 epochs or so we seem to reach the intrinsic error of our methodology.

Figure 7.9: Normalized Accuracy after 100 Epochs

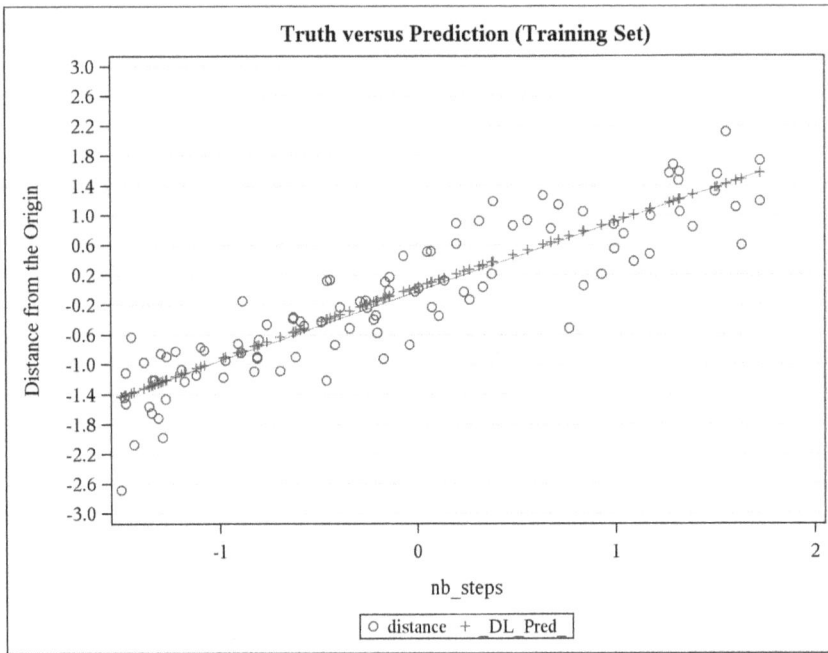

Note that we need fewer epochs for the training than we needed in Chapter 3, because, as we stated earlier, the function is easier to learn. The code in `dl_rw_regression.sas` prints the results of the `deepLearn.dlScore` action that you can see in Figure 7.10. You might be surprised to read 80,000 observations and not 100,000, but remember that we use 20,000 observations for the validation data set.

Figure 7.10: dlScore after 100 Epochs during Normalized Training

Score Information for TRAINING	
Number of Observations Read	80000
Number of Observations Used	80000
Mean Squared Error	0.178789
Loss Error	0.089394

Output CAS Tables			
CAS Library	Name	Number of Rows	Number of Columns
CASUSER(dl4na)	function_scoring	80000	4

It seems appropriate to note one fact at this point: we were able to define and train a brand new DNN with a few lines of code. This is mostly due to the power of DL and to the DL framework that we developed in

Chapter 3. We will see shortly that the power of DL readily transfers to much more complicated problems than the distance from the origin in a random walk.

Now that we have a trained neural network, let's see how well it does with brand new data, our testing set.

Inference Using the Network

The inference code will look familiar if you recall what we did in Chapter 3. The main differences are related to normalization and plotting.

Let's have a quick look at the code:

Program 7.14: Inference Code

```
options cashost='fsnlax05' casport=5570;
cas mysession;

libname DL_LIB cas sessref = mysession;

/* Load the scoring data into CAS */
proc import datafile='~/testing_data.csv'
            out=function_testing;
run;

/* Limit the scoring to 10K values (since this is
   what we measured with the other methodologies) */
proc surveyselect data=function_testing
   method=srs n=10000 out=function_testing noprint;
run;
```

After logging in to CAS, we take a sample of the test data that we generated earlier using the GPU devices on our machine.

Next, we need to normalize the test data, since we computed the weights and biases of our DNN in a normalized space. We start by loading the normalization parameters (mean and standard deviation) into CAS:

Program 7.15: Loading Normalization Parameters

```
proc cas; /* load normalization parameters */
  table.loadTable/
    casOut = "full_function_training_stat"
    path   = "full_function_training_stat.sashdat";
quit;
```

We load those normalization parameters into SAS macro variables and then normalize the 10,000 observations that we want to score:

Program 7.16: Normalizing the Observations

```
data _null_; /* Load normalization parameters into macros */
  set DL_LIB.full_function_training_stat;
  put _name_ ' ' mean ' ' std;
  call symput(cats(_name_, '_mean'), mean);
  call symput(cats(_name_, '_std'), std);
run;

data DL_LIB.function_testing;
  set function_testing;
  nb_steps = (nb_steps - &nb_steps_mean.) / &nb_steps_std.;
  distance = (distance - &distance_mean.) / &distance_std.;
  limit    = (limit - &limit_mean.) / &limit_std.;
run;
```

Because `DL_LIB` is defined as a caslib, the results (`DL_LIB.function_testing`) are in CAS, ready for scoring, which is our next step:

Program 7.17: Scoring the Test Data

```
proc cas;
%include "~/model_io.sas";  /* load_model, etc. */
%include "~/ua.sas";        /* inference, train, etc. */

model_name    = "distance_from_origin_in_rw_model";
caslib_name   = "CASUSER(dl4na)";
testing_table = "function_testing";
scoring_table = "function_scoring";

/* load the model that was previously trained */
load_model(caslib_name, model_name);

/* Now let's score the test set */
copy_vars = { "nb_steps", "distance", "limit" };
inference(model_name, testing_table, scoring_table, copy_vars);

quit; /* proc cas */
```

The mechanics of these procedures are identical to what we did in Chapter 3, except for the table and variable names. Here again, our DL framework comes in very handy. At this point, the test data has been scored, so we can present the results. First, we select only a few to print:

Program 7.18: Selecting Data to Print

```
proc surveyselect data=DL_LIB.function_scoring
   method=srs n=20 out=function_scoring noprint;
run;

data function_scoring;
  set function_scoring;
  nb_steps = nb_steps * &nb_steps_std. + &nb_steps_mean.;
  distance = distance * &distance_std. + &distance_mean.;
  limit    = limit * &limit_std. + &limit_mean.;
  _DL_Pred_= _DL_Pred_ * &distance_std. + &distance_mean.;
run;
proc print data=function_scoring; run;
```

We de-normalize the data that we want to print so that it looks like what one would expect for variables such as a number of steps or a distance.

Next, we compute the difference between the computed distances and the theoretical limit, because we hope to have improved the value of that difference by going to the DNN:

Program 7.19: Computing the Difference between Computed Distances and Theoretical Limit

```
data function_scoring;
  set function_scoring;
  difference      = abs(limit - distance);
  difference_pred = abs(limit - _DL_Pred_);
run;

proc summary data=function_scoring nway;
  var difference difference_pred;
  output out=function_scoring_summary (drop=_:) mean=;
run;

title
  'mean(abs(limit - distance)) and mean(abs(limit - _DL_PRED_))';
proc print data=function_scoring_summary;
run;
title;
```

Finally, we plot the distance values from the test data and from our predictions, side by side:

Program 7.20: Plotting Distance Values

```
ods graphics / reset width=6.4in height=4.8in imagemap attrpriority=none;
proc surveyselect data=DL_LIB.function_scoring
    method=srs n=200 out=function_scoring noprint;
run;

proc sgplot data=function_scoring;
  title "Truth versus Prediction";
  YAXIS LABEL = 'Normalized distances from the Origin in RW'
      GRID VALUES = (-3 TO 3 BY 0.4);
  scatter x=nb_steps y=distance;
  scatter x=nb_steps y=_DL_Pred_ / ;
run;
ods graphics / reset;
```

And with this, we're done.

The first results that we are interested in are the results of dlScore. Does our network generalize well to data that it has never seen? As you can see in Figure 7.11, it turns out that the answer is positive. We observed a loss on the training data set of 0.089394, and we now observe a loss of 0.089721.

Figure 7.11: dlScore on Test Data

Score Information for FUNCTION_TESTING	
Number of Observations Read	10000
Number of Observations Used	10000
Mean Squared Error	0.179442
Loss Error	0.089721

Output CAS Tables			
CAS Library	**Name**	**Number of Rows**	**Number of Columns**
CASUSER(dl4na)	function_scoring	10000	4

Next, we want to have a look at the predictions that our DNN made. You can review a sample of those in Figure 7.12. If you look at the values, you can see that they are not surprising. All values are close to the theoretical limit and not that far from the distances estimated by Monte Carlo simulations.

Figure 7.12: Predictions on Test Data

Obs	nb_steps	distance	limit	_DL_Pred_
1	315809	460.128	448.3863	447.50286097
2	320177	475.84	451.4765	450.51201663
3	321175	436.244	452.1796	451.19951557
4	325422	432.044	455.1594	454.12537466
5	342364	460.912	466.8572	465.79667605
6	364727	501.68	481.8635	481.20370156
7	367500	471.784	483.6918	483.11474689
8	372184	499.368	486.7645	486.34119745
9	376724	491.116	489.7244	489.46334308
10	389327	516.596	497.8486	498.09448616
11	402529	486.108	506.2192	506.22770261
12	406382	499.908	508.6362	508.60700405
13	416835	503.62	515.1363	515.14924059
14	425526	536.784	520.4789	520.57861641

Obs	nb_steps	distance	limit	_DL_Pred_
15	435117	535.94	526.3118	526.57028145
16	453858	532.364	537.5267	538.27579093
17	458256	580.5	540.1248	541.01876305
18	475221	541.052	550.0319	551.59677822
19	489825	524.284	558.4194	560.70259127
20	489867	544.564	558.4434	560.72879813

In Figure 7.13, we can see that the good results that we just saw are confirmed graphically.

Figure 7.13: Accuracy on Test Data

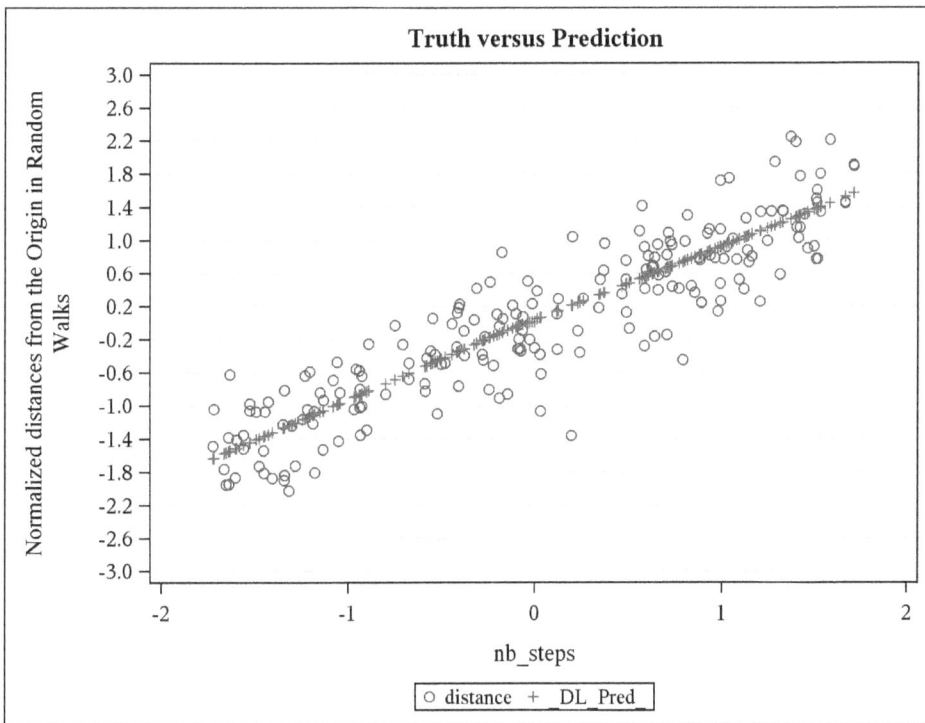

When we loaded the training data, we noticed the differences between the distances computed via Monte Carlo simulations and the theoretical limits. At the time, we hypothesized that the DNN might improve that measure of accuracy (presumably because the DNN would "average" the results of multiple Monte Carlo simulations). In Figure 7.14, you can see that we reduced that difference by more than an order of magnitude. When you run the scoring with your hardware, chances are that you might see different results, but the improvements were always multiples in all the tests that we ran.

Figure 7.14: mean(abs(limit – distance)) and mean(abs(limit – _DL_PRED_))

Obs	difference	difference_pred
1	15.7195	0.76375

The fact that we have good accuracy with the DNN shouldn't obscure our initial goal: to significantly improve the computation of the distance from the origin in random walks. Let's look at our results.

Performance Summary

In Table 7.2, we summarize the performance numbers that we obtained for estimating the distance from the origin in random walks.

As you recall, in Chapter 5, we computed the estimations with Monte Carlo simulations. First, we ran single-threaded Monte Carlo simulations with SAS. Then we leveraged a many-task computing (MTC) framework (SAS Infrastructure for Risk Management) to run the simulations in parallel. The first two lines of Table 7.2 summarize the performance for the Monte Carlo simulations. The improvements of parallelization are quite impressive: we went from approximately 39 hours to 1 hour.

In Chapter 6, we introduced hybrid architectures by adding a couple of GPU devices. That was another appreciable improvement in performance with 32 seconds. Alas, that improvement came at a significant cost. The GPU device was inexpensive, but the programming investment in CUDA was significant. In this chapter, we introduced DNNs and got the best performance with 6 seconds (and no CUDA programming). Both the CUDA version and the DNN version take advantage of the GPUs. But in the DNN case, the performance improvement was free for us as SAS developers, since we leveraged the CAS implementation of DNN.

We need to point out a few caveats about the last two lines of Table 7.2:

- Even though the vast majority of the computing is done on the GPU, some computing is still done on the CPU. This is true for pure CUDA programming and for DNN.
- It is easy to modify the improvements in favor of the DNN. By increasing the number of steps in the random walks by a factor of 10, all the durations in Tables 3 would increase by an order of magnitude, except the 6 seconds. No matter how many steps in the random walk (that is, how complicated the model), the DNN inference will always be 6 seconds.
- Finally, we should point out that the performance improvements given to us by the DNN are more important than what Table 7.2 shows. That is because hidden in the 6 seconds used to perform the inference is a bit less than 2 seconds spent loading the DNN in CPU and then GPU memory. We will revisit this discussion on latency in Chapter 9.

Table 7.2: Distance from Origin in Random Walks (10K)

Chip	Hardware	Technology	Time in Seconds
CPU	60 HT Intel E5-2698 v3 @ 2.30Ghz 500 GB RAM	Single-Threaded	141,400
		Multi-Threaded	4,393
GPU	NVIDIA K80 4992 CUDA Cores @ 840Mhz	CUDA	32
		DNN	6

Other Examples

In this section, we take a cursory look at other examples of deep learning for numerical applications (DL4NA) where the use of a DNN has significantly improved the performance of analytics. Those examples are in the financial sector.

We first look at the pricing of American-style put options.

Pricing of American Options

A put option is a contract that gives its holder the right to sell some asset at a given price. For example, you might own an option that gives you the right to sell an NVIDIA common share for $500. The NVIDIA common share is called the underlying asset, and the $500 is called the strike price. The option also has a maturity date. For example, continuing with this example, with a maturity of 1 year for an American put option on an NVIDIA common share with a strike price of $500, the holder of the option has the right to sell that NVIDIA common share for $500 at any time between the time of purchase and 1 year later.

There is also a call option that gives its owner the right to purchase the underlying asset.

You would buy a put option if you believed that the price of the asset will go down before the maturity date. If the NVIDIA stock plunges to $100 and you own a put option with a strike price of $500, you are going to make $400 on the transaction (things are in fact more complicated because of the cost of the transaction and the cost of the option).

The important thing to remember for this discussion is that since there is a non-negligible amount of money at stake, the key question when it comes to an option is the price: What is a fair price for that option?

The function that we need to estimate for the pricing of American put options is more complex than the distance from the origin in a random walk. It is more complex because of the number of arguments of the function and also because of the implementation of the pricing algorithm. Let's look at the arguments of the function to give us a better idea of the complexity involved in this example.

The function for the pricing of American put options can be summarized as follows:

price = f(strike_price, yield_rate, asset_price, volatility, time_to_expiration, risk_free_rate)

The variables in this function are as follows:

strike_price
    is the price to be paid for the asset at exercise time (the $500 in our example).

yield_rate
    Is the yield of the asset. In our example, that would be the rate of the dividends paid by NVIDIA to its stock holders.

asset_price
    is the price of the asset at the time of the purchase of the option. In our example, it would be the price of an NVIDIA common share at the time of the pricing of the option.

volatility
    is a measure of the volatility of the asset, in this case the volatility of NVIDIA common stocks.

time_to_expiration
> is the time to the maturity of the option. In our example, this would be the number of days left until the one-year maturity.

risk_free_rate
> is the rate of a very safe investment (for example, US government treasuries).

There is no closed-form solution to this function at this time, so we typically need to resort to simulations to estimate the value of the function. A detailed description of the implementation of the pricing algorithm is outside of the scope of this book, since it is not related to DL, but it can be found in Orosi (2015).

More relevant to the topic of this book, Izquierdo (2017) explains how to approximate the complex pricing of American put options using a DNN. In that paper, using a methodology almost identical to what we outlined earlier in this chapter, the author compares the performances of different pricing algorithms. Unsurprisingly to us (based on what we did with random walks), the findings summarized in Table 7.3 shows a significant performance gain in favor of the DNN.

Table 7.3: Pricing American Put Options

Chip	Hardware	Technology	Time in Seconds
CPU	60 HT Intel E5-2698 v3 @ 2.30Ghz 500 GB RAM	Single-Threaded	11,242
		Multi-Threaded	565
GPU	NVIDIA K80 4992 CUDA Cores @ 840Mhz	CUDA	56
		DNN	13

Using the DNN turned out to be about 4 times faster than the pure CUDA programming. But the real story is that between the SAS program and the DNN, we see that the DNN is 860 times faster than the single-threaded version and 43 times faster than the multi-threaded version. These time improvements are the real story, because you can get this level of performance improvement by writing SAS code. No CUDA coding is required on your part, because you leverage the CUDA programming done in CAS.

This last statement highlights a key advantage of DL4NA. Under some conditions (which we discussed in Chapter 3), you can, using only SAS programming and inexpensive hardware, gain orders of magnitude of performance improvements for your SAS programs.

Let's now have a look at another example: the pricing of complex insurance products. As you will see shortly, DL4NA also provides a significant performance improvement in this example.

Pricing of Variable Annuities Contracts

A variable annuity is a contract between a policyholder and an insurance company. The buyer of the contract (or policyholder) can accumulate assets in the account of the variable annuity via investments in mutual funds. The added value of the variable annuity comes in the form of guarantees that help protect the buyer from market downturns like the dot-com bust or the financial crisis that preceded the great recession. Those guarantees are typically in the form of guaranteed minimum death benefits (GMDB), guaranteed minimum accumulation benefits (GMAB), guaranteed minimum income benefits (GMIB), guaranteed minimum withdrawal benefits (GMWB), or the combination of both GMDB + GMWB.

As you might imagine, the availability of those guarantees is a risk to the insurance companies: How would they pay those guaranteed benefits in case of adverse events such as a market crash or even an industry sector crash (for example, the crash of the LISP machines market that we discussed in Chapter 2)? A popular risk-mitigating technique is dynamic hedging. The insurance company buys portfolios of financial derivatives such as the American put options that we just discussed in the hope that their payoffs will offset the payouts of the guarantees to the policyholders. With the guarantees linked to mutual funds, the portfolio must be constantly evaluated to keep the exposure or the risk up-to-date. Traditionally, this evaluation is done using Monte Carlo simulations. As we know very well by now, it is possible to significantly improve the performance of Monte Carlo simulations using DL, and that's exactly what Chen and Bequet (2017) describe.

The function that we need to estimate for the pricing of variable annuities is the most complex that we have looked at so far. The number of arguments is comparable to what is required for American options, but the model itself is more complicated (the GPU implementation is hundreds of lines of CUDA code).

The function for the pricing of variable annuities can be summarized as follows:

price = f(risk_free_rate, mortality_rate, account_value, maturity, age, gender,
guarantee_type, withdrawal_rate)

The variables in this function are as follows:

risk_free_rate
is the rate of a very safe investment (for example, US government treasuries).

mortality_rate
is a value based on the mortality table, which is used by actuaries to determine the probability of death in a population characterized by gender and age.

account_value
is the value of the mutual fund account associated with the variable annuity.

maturity
is the number of years to the maturity of the contract.

age
is the age of the policy holder.

gender
is the gender of the policy holder.

guarantee_type
is the type of guarantee associated with the contract: GMDB, GMWB, or both.

withdrawal_rate
is the annual rate at which the policy holder will withdraw money from the mutual fund account.

In addition to these arguments, the pricing algorithm must consider multiple market scenarios (whether the market is up or down, and by how much).

The numbers in Table 7.4 report DNN performance improvements that are better than what we measured for the distance from the origin in random walks. This is because more computations were needed for each simulation with variable annuities, while the computations needed for inference are always the same. This last fact is true because the architecture of the network is the same (8 layers of 1,024 ReLU neurons).

Table 7.4: Pricing American Variable Annuities

Chip	Hardware	Technology	Time in Seconds
CPU	60 HT Intel E5-2698 v3 @ 2.30Ghz 500 GB RAM	Single-Threaded	NA
		Multi-Threaded	2,640
GPU	NVIDIA K80 4992 CUDA Cores @ 840Mhz	CUDA	52
		DNN	4

Conclusion

We started this chapter by leveraging GPUs to generate training, validation, and test data. We then trained a DNN to approximate the distance from the origin in random walks. Once we had normalized the data, the training of the DNN gave us no troubles, thanks to the DL framework that we developed in Chapter 2.

Armed with a trained DNN, we ran through the inference on our test data and noticed that the performance of our analytics was orders of magnitude faster than our multi-threaded implementation. The use of the dlScore action enabled us to achieve these significant performance improvements without writing a single line of CUDA code.

We then briefly looked at more complicated models such as the American put option and the variable annuity contract, where the performance gains are even greater than what we could achieve for random walks of an average of 400,000 steps or so.

All the operations that we performed in this chapter were focused on a single user and a single run. In the next chapter, we will leverage MTC to package our analytics so that they can be deployed in an enterprise where many users need to contribute to and share analytics.

Chapter 8: Deep Learning for Numerical Applications in the Enterprise

In this chapter, we package the code that we have developed in the previous chapter using an enterprise-class many-task computing (MTC) platform (SAS Infrastructure for Risk Management). We make extensive use of the concepts that we presented in Chapter 4 and Chapter 5, so it might be a good idea to make sure that you are familiar with the material in those chapters before reading this one.

Enterprise Applications

The main goal of an enterprise application is to satisfy the needs of an organization—in other words, the needs of many people. This purpose is in contrast with what we've done so far in this book. We've been focusing on the need of one person to develop simulations, train deep neural networks (DNN), run scoring, and so on. In this chapter, we focus on the SAS developer as part of an organization consisting of other SAS developers and users who are not necessarily SAS developers.

To serve the needs of an organization, many software vendors rely on automation of the business processes (Fowler 2003). We will use job flows in SAS Infrastructure for Risk Management to document and automate the computational aspects of the business processes that we implement.

This chapter is organized around the workflow of a SAS developer who wants to develop and share analytics powered by deep learning (DL). Specifically, the end goal of this chapter is to provide all users of the enterprise with a job flow that automates the following workflow:

1. Extract, transform, and load (ETL) the training and test data.
2. Normalize the training and testing data.
3. Train the DNN.
4. Score the testing data with the DNN.

As you will see as we progress, this workflow involves many technologies, from SAS 9.4 to SAS Viya to CAS to DL running on CPUs and GPUs. Our MTC framework makes integrating these technologies easy.

We use SAS Studio as an integrated development environment (IDE), but most of the work that we describe here can be performed by any SAS client. There is some overlap between this chapter and Chapter 5 in the mechanics of using SAS Studio, but rather than constantly referring you back to Chapter 5, we thought that this chapter would be easier to read if it stood by itself as much as possible.

Since we have decided to leverage MTC, we start our implementation by developing a task.

A Task

The first step after installing SAS Infrastructure for Risk Management and SAS Studio on your machine is to set up your development environment as described in Appendix A. After this is done, you should see something like Figure 8.1.

Figure 8.1: SAS Studio

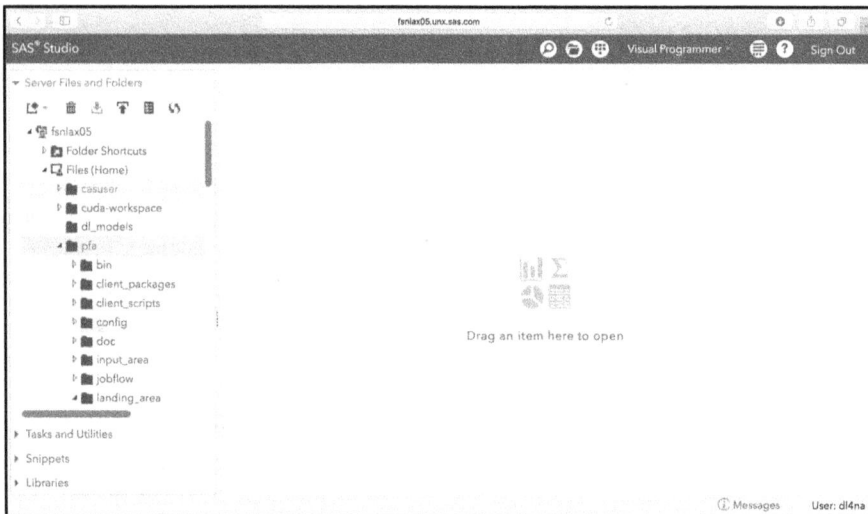

The `pfa` folder is your personal federated area (PFA). All the artifacts that we develop in this chapter will be in the `pfa` folder, but we will export them to a federated area (FA) so that we can share them with other users.

Data

Let's start with loading the training data that we need into the `landing_area` folder.

We begin by creating an empty folder in the `landing_area` folder. To do that, scroll down the list of folders under the `pfa` folder, right-click on the `landing_area` folder, and select **New**, followed by **Folder**, as you can see in Figure 8.2.

Figure 8.2: Creating a New Folder in the landing_area Folder

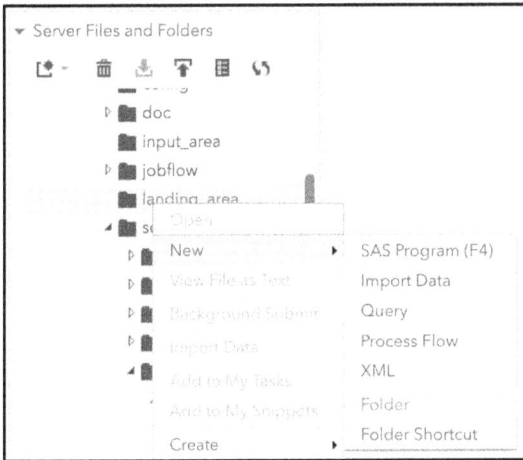

A dialog box appears so that you can enter the new folder name (`training`), as shown in Figure 8.3.

Figure 8.3: New training Folder

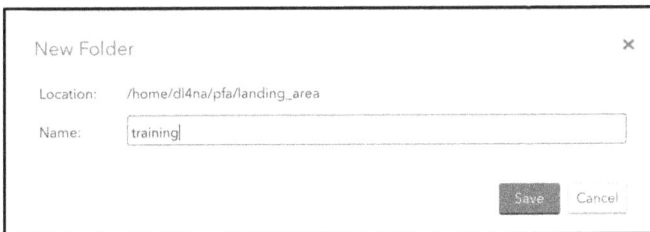

Now that we have a folder to host the data, let's create a SAS program to populate it. To achieve that goal, click the upper left icon and select **SAS Program (F4)**, as shown in Figure 8.4.

Figure 8.4: New SAS Program

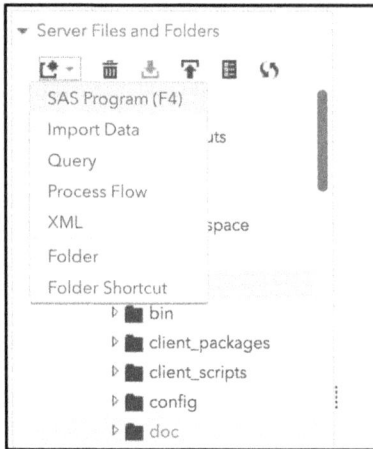

This gives you a blank SAS file in which you should enter the following code to import the CSV file into a SAS7BDAT file:

Program 8.1: Loading the Training Data Set

```
libname trndata '~/pfa/landing_area/training';

/* Load the training data set into CAS */
proc import datafile='~/training_data.csv'
            out=trndata.full_function_training;
run;
```

We also need a mapping in the `landing_area` folder so that the tasks can use the training data:

Program 8.2: Defining Librefs for the Task

```
/* Define a libref mapping for the task */
data _null_ ;
    FILE "~/pfa/config/libnames.txt";
    PUT 'trndata=%la/training';
run;

/* We also need to touch last_update.txt to trigger live ETL
   and a refresh of the librefs */
data _null_ ;
    FILE "~/pfa/input_area/last_update.txt";
    PUT;
run ;
```

You might recall that in Chapter 5, we relied on the SAS Infrastructure for Risk Management live ETL feature to load input data for job flows. When developing a new flow definition with new data in your PFA, there is no risk of colliding with other users, so we can directly load the data into `landing_area` (in other words, we can skip the `input_area`). However, we still need to make sure that `libname.txt` is up-to-date with the librefs used in the inputs of tasks, hence the last DATA step that triggers live ETL.

Another technicality to keep in mind is that the code that we just entered is executed as the logged-in user (dl4na in our case), while the tasks will be executed as the SAS server user, which by default is sassrv. This piece of information will come in handy should you run into permission problems on the file system.

Save the file to the **client_scripts** folder of your PFA so that the situation looks like Figure 8.5.

Figure 8.5: etl_training.sas

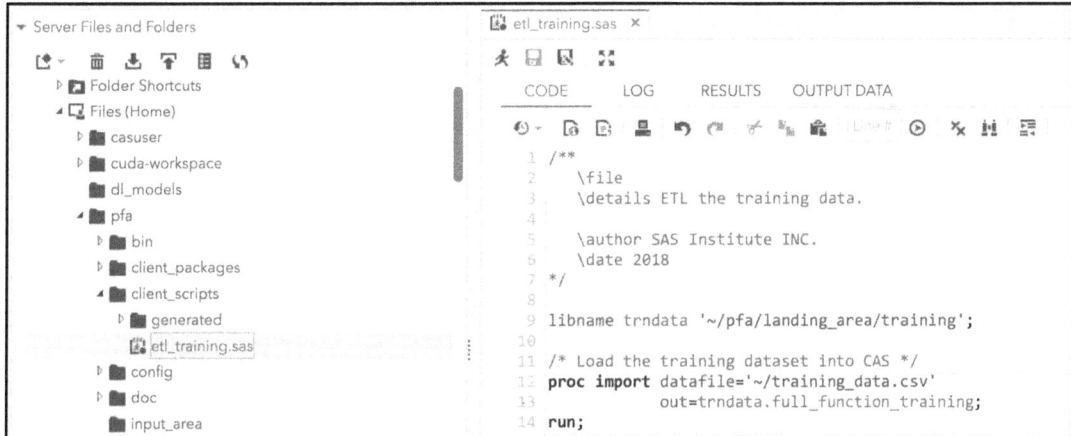

Run the ETL code by clicking the running figure icon (🏃).

After execution, you should have your training data loaded into the **landing_area** folder as shown in Figure 8.6 (double-click full_function_training.sas7bdat to review the file content in SAS Studio). As expected, we have 1,000,000 observations in full_function_training.sas7bdat.

Figure 8.6: Training Data

Now that we have training data, we're ready to code the implementation of our task.

Task Implementation

You might recall that in the previous chapter we normalized the training data in order to improve the training. In that process, we relied on two SAS Infrastructure for Risk Management macros: `%irm_normalize_data` and `%irm_word_count`. So let's copy those two files from the code folder of this chapter to the `~/pfa/source/sas/ucmacros` folder so that the situation looks like Figure 8.7.

Figure 8.7: SAS Infrastructure for Risk Management Macros

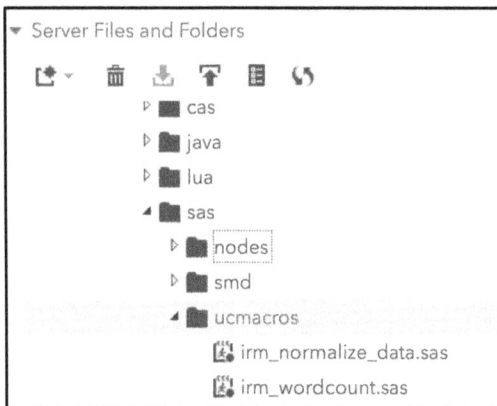

As you did previously, create a new empty SAS program and type or copy and paste the following code:

Program 8.3: Preparing the Training Data

```
/**
   \file
   \brief   Prepare training data.
   \details A task to take a subset of our pool of data for
            training of a DNN that approximates the distance
            from the origin in random walks.
            This task also computes the difference
            between the theoretical limit and the distance.
            Finally, this task normalizes the data for better
            conversion during the training.
            Normalize the training data for a DNN that
            approximates the distance from the origin in
            random walks.

   \param[in]  %input=TRNDATA.full_function_training.sas7bdat
                       Training data.
   \param[out] %output=NORMDAT.full_function_training.sas7bdat
                       Normalized training data.
   \param[out] %out_stat=NORMDAT.full_function_training_stat.sas7bdat
                       Normalization stats.

   \author SAS Institute INC.
   \date 2018
*/
```

The header is used to generate the documentation and sequence the tasks in the job flow.

We start the implementation of our task by limiting the training set to 100,000 observations, since we know from the previous chapter that this is plenty to learn the function at hand. Notice the use of &input. In the following code, this is how tasks support substitution in their input or output arguments:

Program 8.4: Limiting the Training Set

```
/* Limit the training set */
proc surveyselect data=&input.
   method=srs n=100000 out=function_training;
run;

data function_training;
  set function_training;
  difference = abs(limit - distance);
run;
```

Finally, we normalize the data as we discussed previously:

Program 8.5: Normalizing the Data

```
%irm_normalize_data(in_tbl       =function_training,
                    var_lst       =nb_steps distance limit,
                    out_tbl_stat =&out_stat.,
                    out_tbl       =&output);
```

We then save the code to **~/pfa/source/sas/nodes** so that the situation on your machine looks like Figure 8.8.

As you can see, the SAS code in the task doesn't contain any synchronization to allow tasks to run in parallel. Our MTC platform will do the hard work for us.

Figure 8.8: Prepare Training Data

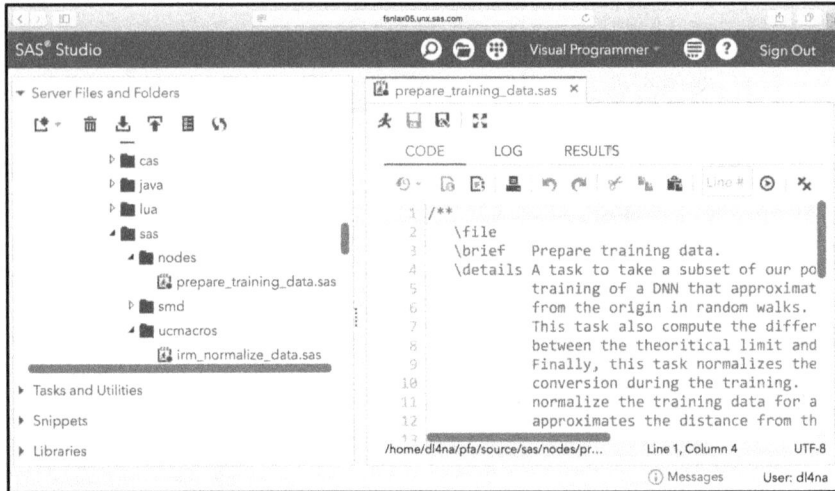

To run a task, we need a flow instance, so let's use the scripting client to perform those tasks.

A Simple Flow

As we did before, we create an empty SAS program in SAS Studio and save it in **~/pfa/client_scripts** (with `etl_training.sas`) under the name `prepare_data_flow`.

The code is straightforward. After initializing the SAS Infrastructure for Risk Management scripting client with the `%irm_sc_init` macro, we create the flow definition (and name it `prepare_data_flow`):

Program 8.6: Building the Job Flow

```
/* Build the job flow (definition) */
%irm_sc_build_jobflow(
        i_jf_name      =prepare_data_flow,
        o_jf_ref_name =prepare_data_flow_ref);
```

The job flow contains only one task, the one we just created in SAS Studio:

```
/* Add one task */
%prepare_training_data(i_jf_ref =&prepare_data_flow_ref);
```

As you have seen before, SAS Infrastructure for Risk Management automatically picks up your change in SAS Studio (in this case the creation of a new task) and automatically generates a SAS macro so that you can easily add your new task to a job flow.

After saving the job flow definition, we can create and execute an instance of that job flow definition:

Program 8.7: Executing the Job Flow

```
/* Save job flow definition to SAS Infrastructure for Risk Management Server */
%irm_sc_save_jobflow(
        i_jf_ref =&prepare_data_flow_ref);

/* Execute the job flow on the SAS Infrastructure for Risk Management Server */
%irm_sc_execute_jobflow(
        i_jf_ref =&prepare_data_flow_ref);
```

After the execution of `prepare_data_flow.sas`, you should see something like Figure 8.9 in the SAS Infrastructure for Risk Management GUI (for example, `http://fsnlax05:7980/SASIRM`).

Figure 8.9: The prepare_data_flow Job Flow Instance

If you drill down into the `prepare_data_flow` instance, you will see something like Figure 8.10.

Figure 8.10: prepare_training_data in prepare_data_flow

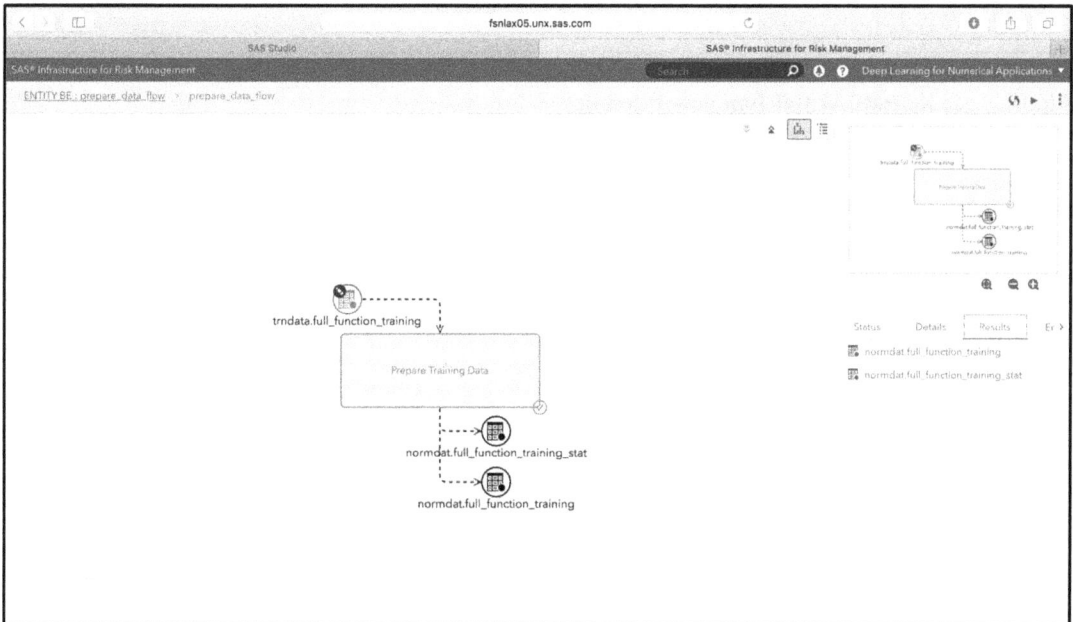

In Figure 8.10, you can see the input data that we loaded into the **landing_area** folder and the output of the task that we just wrote. You can either view the outputs in Microsoft Excel (double-click the circle) or view the outputs in SAS Studio (right-click the circle and select **View in SAS Studio**).

Figure 8.11 shows the normalization statistics in SAS Studio, and Figure 8.12 shows the normalized training data in SAS Studio as well. There are no surprises here; this is what we would expect. (See Chapter 7 for a thorough discussion of the normalization data and statistics).

Figure 8.11: NORMDAT.full_function_training_stat

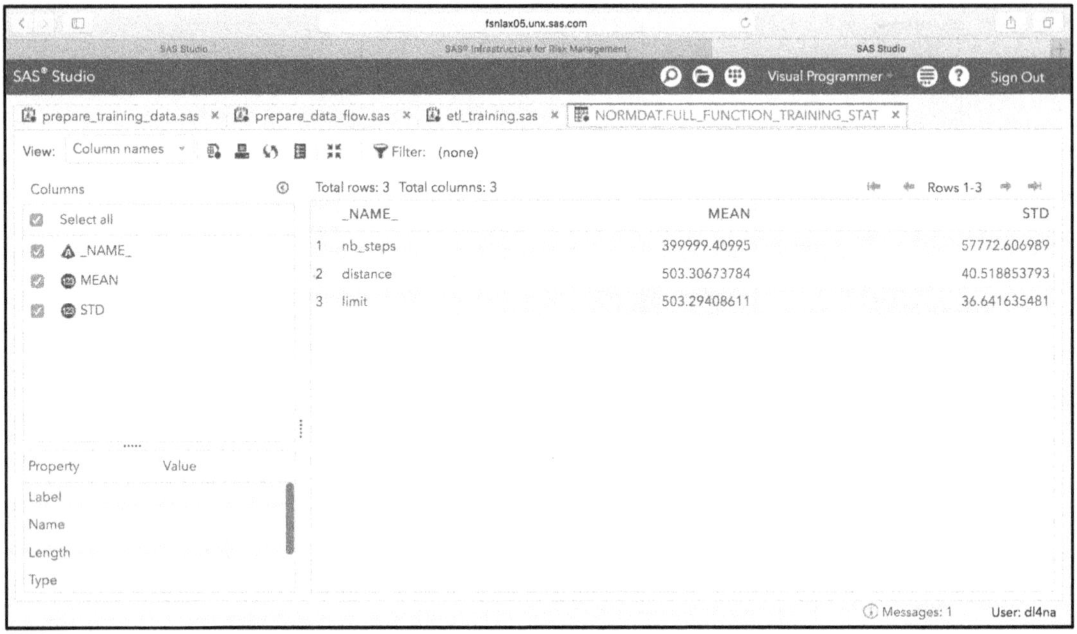

Figure 8.12: NORMDAT.full_function_training

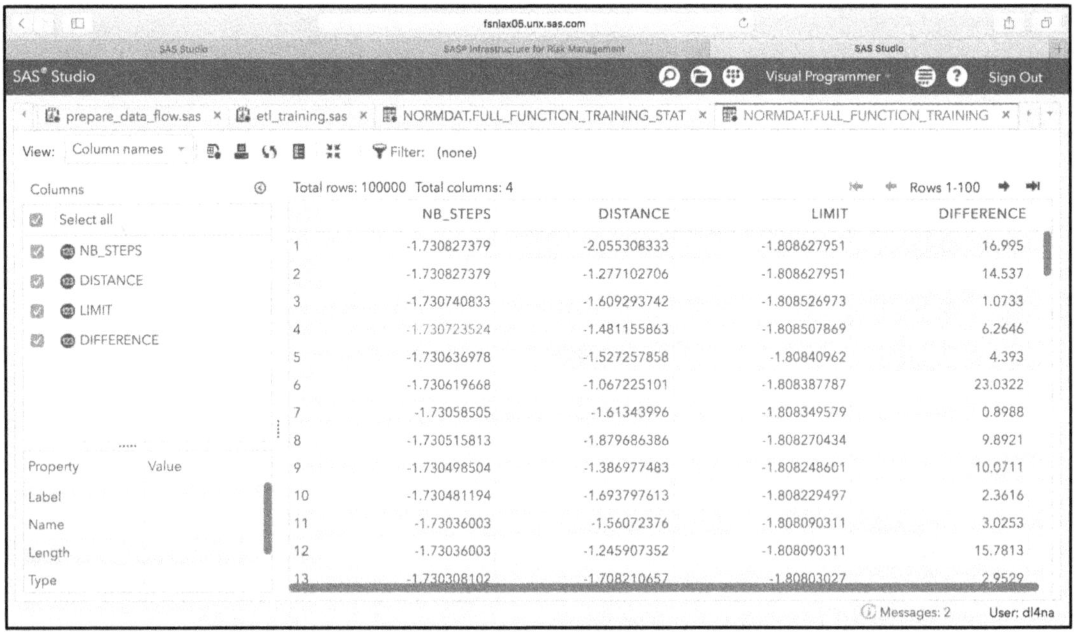

We now need to repeat the same process for the testing data. After ETL, things should look like Figure 8.13, and after adding a task for the testing data to the `prepare_data_flow` job flow definition, the resulting flow should look like Figure 8.14. Notice the parallel execution of the preparation for the training and testing data.

Figure 8.13: function_testing.sas7bdat

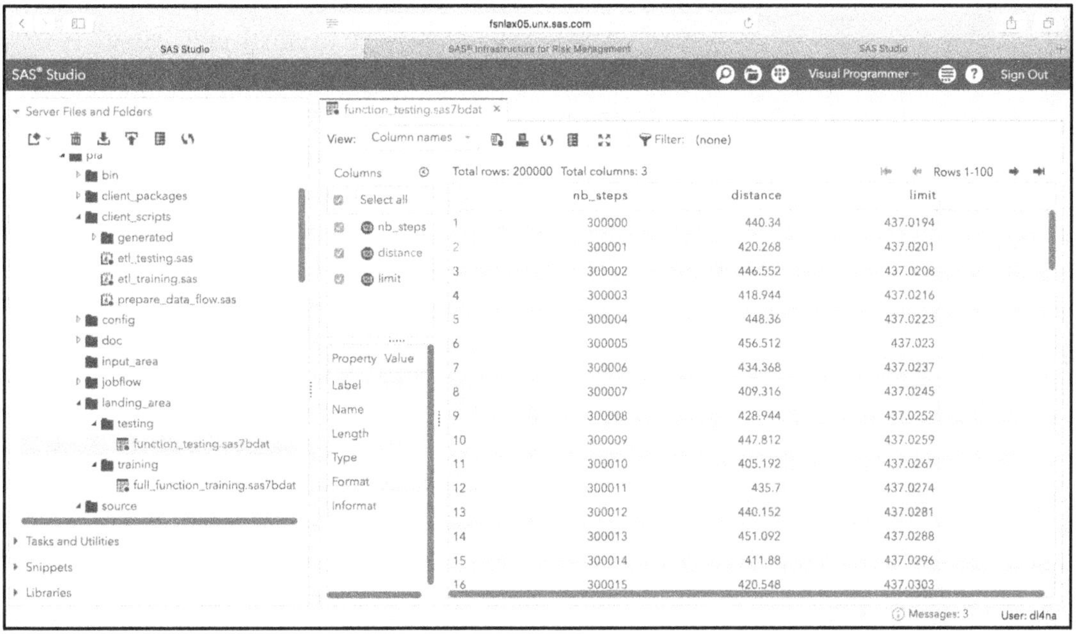

	nb_steps	distance	limit
1	300000	440.34	437.0194
2	300001	420.268	437.0201
3	300002	446.552	437.0208
4	300003	418.944	437.0216
5	300004	448.36	437.0223
6	300005	456.512	437.023
7	300006	434.368	437.0237
8	300007	409.316	437.0245
9	300008	428.944	437.0252
10	300009	447.812	437.0259
11	300010	405.192	437.0267
12	300011	435.7	437.0274
13	300012	440.152	437.0281
14	300013	451.092	437.0288
15	300014	411.88	437.0296
16	300015	420.548	437.0303

Figure 8.14: Prepare Training and Testing Data

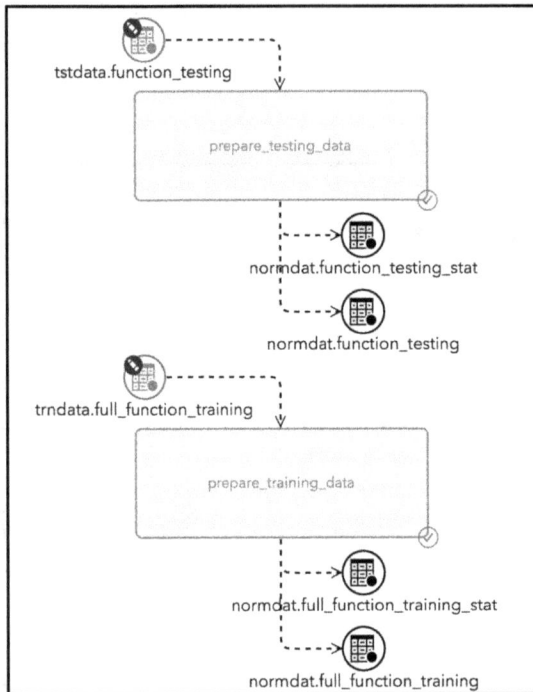

Now that we have training and testing data, we can turn our attention to the definition and training of the DNN.

A Training Flow Task

In Chapter 7, we bundled the definition and the training of the DNN into one task. For simplicity, we will keep this bundling in this chapter, but we should briefly discuss a couple of alternate designs.

As a first alternative we could consider two tasks: one to define the DNN and one to train the DNN. This split would make sense if the creation of the DNN were more complex than the simple fully connected DNN that we have worked with so far. In our case, we only work with one DNN architecture, so one task will do the job.

Another possible design would be to create separate two job flows to define and train the DNNs (not just two different tasks). This alternative makes more and more sense as the number of DNNs and the number of SAS developers increase.

The persistence of the DNN definitions and weights is also a design decision that we face. In the previous chapter, everything went into the user caslib. That choice is unlikely to be ideal in an enterprise environment for several reasons, collisions between multiple users and permissions being the obvious problems. Consequently, we will use a caslib different from the user caslib as a model repository. Specifically, we will use the model repository to save the model definition, and the flow itself to save the weights and biases of the trained DNN. This separation makes sense in our case, because all the regressions and inferences that we perform use the same DNN architecture, but the weights and biases differ from flow instance to flow instance (since we are running training and inference with sets of data that are specific to the flow instance).

Now that we have made some design decisions, we can get started with the implementation. As we did earlier, we create a new task in SAS Studio (rw_regression.sas) that we save in **~/pfa/source/sas/nodes**.

The code is like what we did in the previous chapter, but there are some key (though subtle) differences that we will point out as we review the code.

The first thing to remember when we write a task is that we need to declare our inputs and outputs:

Program 8.8: Declaring Inputs and Outputs

```
/**
   \file
   \brief    Define and train a DNN.
   \details Definition and training of a DNN that approximates
            the distance from the origin in random walks.

   \param[in]   NORMDAT.full_function_training.sas7bdat
                     Training data.
   \param[out] MODEL.rw_dl_weights.sas7bdat
                     DNN weights.

   \author SAS Institute INC.
   \date 2018
*/
```

We take the training data set as input, and we produce the model weights that we learn during the training of the DNN. The DNN definition along with its parameters are located in our repository. (See Chapter 2 for an explanation of the CAS organization of a DNN model).

We then connect to CAS, where we can create and train our model:

Program 8.9: Connecting to CAS

```
options cashost='localhost' casport=5570;
cas mysession;

libname DL_LIB cas sessref=mysession datalimit=all;

data DL_LIB.function_training;
  set NORMDAT.full_function_training;
run;

proc cas;
%include "~/model_io.sas";   /* load_model, etc. */
%include "~/ua.sas";         /* inference, train, etc. */
```

There are a few caveats here:

1. Because of the technicality of running as `sassrv`, we put `model_io.sas` and `ua.sas` in `sassrv`'s home folder.

2. It is possible to use full pathnames.

3. The data that we transfer from CAS could exceed the default limit, so we set the `datalimit=all` when we create the libref to CAS.

The creation of the model along with the training of the model is a verbatim copy of what we did before, except for the name of the model:

Program 8.10: Building and Training the Model

```
/* We load the data */
training_table   = "function_training";
tables           = setup_training_data({name = training_table}, 80);
training_table   = tables.training;
validation_table = tables.validation;

/* We build the model */
model_name           = "rw_model";
number_of_hidden_layers = 8;
define_model(model_name, number_of_hidden_layers);

/* We train the model */
inputs            = {"nb_steps"};
target            = "distance";
categorical_inputs = { };
use_gpu           = 1;
max_epoch         = 100;
train(model_name, training_table, validation_table,
      inputs, target, categorical_inputs, max_epoch, use_gpu);
```

This is a good time to create our model repository, so let's do that. First we need a caslib:

```
caslib model_repository datasource=(srctype="path") NOTACTIVE
                        path="/local/install/cas_model_repository";
```

The preceding statement defines a new caslib with its persistence on the file system (here a folder on a LINUX machine). The `NOTACTIVE` option is to keep the active caslib unchanged (by default, the statement will change the caslib to `CASUSER("sassrv")`, which is the user caslib). The caslib is called `model_repository`. Armed with that caslib, we now have a place to save our model definition:

Program 8.11: Saving the Model Definition

```
/* We save the new model to our model repository */
cas_library_name = "model_repository";
save_model("casuser", model_name, cas_library_name);
```

The first argument to `save_model` (`"casuser"`) is the caslib where the model currently resides (`casuser` is a shortcut for `CASUSER("sassrv")` in this case). The last argument is the caslib where we want to save the model, our model repository. The changes to `save_model()` are in the `model_io.sas` file included with the code of this chapter.

At this point, we can save our model weights to a SAS table that contains the output of our task:

Program 8.12: Saving the Model Weights

```
data MODEL.rw_dl_weights;
  set DL_LIB.rw_model_weights;
run;

cas mysession terminate;
```

And with this we're done with the creation and training of the DNN task. To run the task, we need a flow, so let's create a flow that includes the data preparation tasks. We put those data preparation tasks in a subflow (you typically use subflows to reduce the clutter or emphasize a high-level design of the overall flow).

We create an empty SAS program as we did before, and we call it rw_training_flow.sas. We save it to the **client_scripts** folder of our PFA.

Since the flow creation script uses concepts that we have reviewed before, we omit most code here.

We create two job flows, one for the top-level flow and one for the subflow:

Program 8.13: Creating Job Flows

```
%irm_sc_build_jobflow(
          i_jf_name      =dl_rw_training,
          o_jf_ref_name  =dl_rw_training_ref);

%irm_sc_build_jobflow(
          i_jf_name      =prepare_data_flow,
          o_jf_ref_name  =prepare_data_flow_ref);
```

We simply add the subflow to the top-level flow along with the rw_regression task:

Program 8.14: Adding the Subflow to the Top-Level Flow

```
/* Add the subflow */
%irm_sc_add_subflow(
          i_jf_ref     =&dl_rw_training_ref,
          i_sub_jf_ref =&prepare_data_flow_ref);

/* Add the training task (to the top-level flow) */
%rw_regression(i_jf_ref =&dl_rw_training_ref);
```

After the execution, you should see something like Figure 8.15 in the SAS Infrastructure for Risk Management GUI. The execution might take a while, depending on your hardware. If you were to drill down inside the subflow, you would see the two data preparation tasks that we defined earlier. Note that you could add additional outputs: the loss values during the training and the inference results of the training or validation sets.

Figure 8.15: A Training Flow

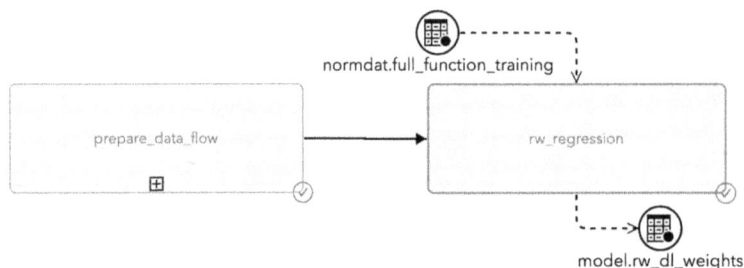

At this point, we have a trained DNN to approximate the distance from the origin in random walks (given a number of steps). To complete the cycle, we need to add an inference task to the top-level flow.

An Inference Flow

We create the inference task as we did before. We call it `rw_score`, and we save it to `~/pfa/source/sas/nodes`.

During the development of the inference task, we need to load the model from the model repository and then apply the weights computed in the training task. Let's add a function to do just that in `model_io.sas`:

Program 8.15 Adding a Function to Load the Model and Apply Weights

```
/*
   This function loads the model of our universal approximator
   from cas_library_name (on disk) into casuser, assuming that
   the weights are already in casuser.
   Loading the model means loading the tables that define the
   model into cas (without forgetting the attributes).

Arguments:
  cas_library_name IN  The name of the cas library
                       to read the model from
  model_name       IN  The name of the model to load
*/
function load_model_in_casuser(cas_library_name, model_name);
  param_table_name      = get_param_table_name(model_name);
  attr_param_table_name = get_attr_param_table_name(model_name);
```

The arguments are identical to the `load_model` function that we developed in Chapter 7, but the implementation is different, since we need to load the tables from one caslib to another (from the model repository to `casuser`). First, we load the model definition from the model repository (passed in `cas_library_name`):

Program 8.16: Loading the Model Definition

```
table.loadTable/
  caslib = cas_library_name
  casOut = { caslib = "casuser" name = model_name }
  path   = model_name || ".sashdat";

table.loadTable/
  caslib = cas_library_name
  casOut = { caslib = "casuser" name = attr_param_table_name }
  path   = attr_param_table_name || ".sashdat";
```

Note that the syntax of the `casOut` argument is logically equivalent to the `libname.table_name` that you would use in traditional SAS programming.

And now we must be sure to apply the attributes to the weights table (see Chapter 3 or the CAS documentation for more information about attribute tables):

Program 8.17: Applying Attributes to the Weights Table

```
table.attribute /
  table   = attr_param_table_name
  caslib  = "casuser"
  name    = param_table_name
  task    = "ADD";
end func;   /* load_model_in_casuser */
```

With the `load_model_in_casuser` function, writing the task for scoring with our model is easy.

As for any task that we've developed so far, we first define our inputs: one table with the test data, one table for the model weights, and one table with the normalization statistics (since we will need to de-normalize the results). We define only one output: the results of the scoring on the test data:

Program 8.18: Defining Inputs and Output

```
/**
   \file
   \brief   Load and score with a DNN.

   \param[in]   NORMDAT.function_testing.sas7bdat
                     Testing data.
   \param[in]   NORMDAT.full_function_training_stat.sas7bdat
                     Normalization stats.
   \param[in]   MODEL.rw_dl_weights.sas7bdat
                     DNN weights.
   \param[out]  RESULT.function_scoring.sas7bdat
                     DNN weights attributes.

   \author SAS Institute INC.
   \date 2018
*/
```

After logging in to CAS (code omitted), we load the testing data and the model weights into the active caslib (`CASUSER(sassrv)` or simply `casuser`). We do not persist the model to this caslib, so the DNN model definitions stay in memory with respect to that caslib.

Program 8.19: Loading Data

```
data DL_LIB.function_testing;
  set NORMDAT.function_testing;
run;

data DL_LIB.rw_model_weights;
  set MODEL.rw_dl_weights;
run;
```

Using our new `load_model_in_casuser` function and our DNN model repository, we can load the model and its weights in memory:

Program 8.20: Loading the Model and Weights

```
proc cas;
%include "~/model_io.sas";
%include "~/ua.sas";

model_name     = "rw_model";
testing_table = "function_testing";
scoring_table = "function_scoring";

/* Load the model from our model repository */
caslib model_repository datasource=(srctype="path") NOTACTIVE
                        path="/local/install/cas_model_repository";
model_name = "rw_model";
cas_library_name = "model_repository";
load_model_in_casuser(cas_library_name, model_name);
```

The scoring code is exactly what we did in the stand-alone case in Chapter 7, including the de-normalization of the test results:

Program 8.21: Scoring the Data

```
/* Now let's score the test set */
copy_vars = { "nb_steps", "distance", "limit" };
inference(model_name, testing_table, scoring_table, copy_vars);
quit; /* proc cas */

/* Select a few observations from the scoring table */
proc surveyselect data=DL_LIB.function_scoring
   method=srs n=20 out=function_scoring noprint;
run;

/* Don't forget to de-normalize the data */
data _null_; /* Load normalization parameters into macros */
  set NORMDAT.full_function_training_stat;
  put _name_ ' ' mean ' ' std;
  call symput(cats(_name_, '_mean'), mean);
  call symput(cats(_name_, '_std'), std);
run;
```

```
data function_scoring;
  set function_scoring;
  nb_steps = nb_steps * &nb_steps_std. + &nb_steps_mean.;
  distance = distance * &distance_std. + &distance_mean.;
  limit    = limit * &limit_std. + &limit_mean.;
  _DL_Pred_= _DL_Pred_ * &distance_std. + &distance_mean.;
run;
```

You might be concerned that the de-normalization of the data could have a multiplicative effect on the error because of the term * &distance_std. That is a valid concern in some applications. In Chen and Bequet (2017), the authors solve the problem by normalizing the inputs, but not the outputs. This type of training is more complicated than what we do for random walks, since it requires a custom error function during the training.

We are now ready to create a complete job flow with data preparation, network training, and scoring. We put the code in **~/pfa/client_scripts/rw_scoring_flow.sas**. The only difference from the training flow is the addition of the scoring task that we just wrote:

```
/* Add the scoring task (to the top level flow) */
%rw_score(i_jf_ref =&dl_rw_scoring_ref);
```

The execution of rw_scoring_flow.sas gives us the result in Figure 8.16.

Figure 8.16: Scoring Flow

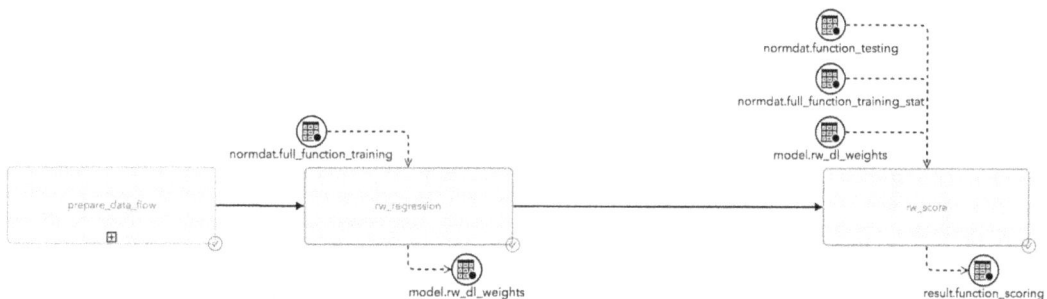

If you run the code yourself, you might notice that the execution literally flies through the rw_regression part instead of doing the training all over again. This is the benefit of data object pooling, which guarantees that tasks previously executed won't execute again. If you were to look in the SAS Infrastructure for Risk Management server log, you would see a log message like this one, confirming that indeed the execution of the expensive training was skipped:

```
com.sas.solutions.risk.irm.server.jobflowexec.JobFlowExecutor  - Skipped
execution of task rw_regression(key: 2096377225, parent flow: dl_rw_scoring,
bpmn: dl_rw_scoring.bpmn) because the output data is available in the data pool.
```

As you might expect, the execution of the scoring task is not skipped:

```
com.sas.solutions.risk.irm.server.pooling.DataObjectPoolingService   - The
execution of task rw_score(key: 750905698) is not skipped because
DataObjectPoolingService did not find its output in the pool.
```

Data object pooling is an essential part of an enterprise deployment of your DL analytics. Without data object pooling, much unnecessary retraining of the DNN would reoccur, and your productivity would suffer greatly. We haven't discussed how to share your DL and MTC work with others yet, but let us mention at this point that if other users were to run the same scoring as we just did, then not only would the training (regression) task be skipped, but also the scoring task. This is another form of machine learning (ML) where the system gets faster and faster as it automatically learns from the inputs, flows, and outputs of all users.

Documentation

If you were to hold your mouse pointer over the tasks defined in our flow, you would get an error message like that in Figure 8.17.

Figure 8.17: No Tooltip

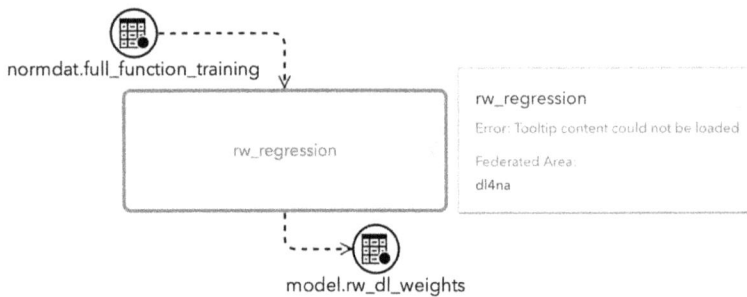

Adding a simple call to the following macro in `rw_scoring_flow.sas` solves the problem:

```
%irm_sc_gen_doc();
```

The tooltip should now look like Figure 8.18. Note that a right-click on the task now displays the documentation menu selection to display the documentation generated from the metadata in the header of our tasks.

Figure 8.18: Tooltip

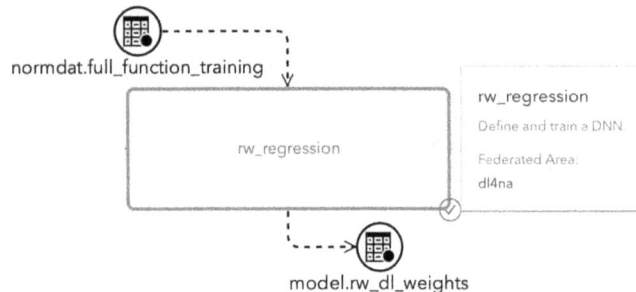

SAS Infrastructure for Risk Management doesn't generate the documentation by default, because it usually takes a few cycles and is not essential during the development phase.

Let's now say a few words about heterogeneous architecture support.

Heterogeneous Architectures

You might not have realized it, but the job flow that we just created runs on a heterogeneous architecture. In other words, the job flow runs on CPUs and GPUs. Other than specifying the `gpu = true` option on the `deepLearn.dlTrain` action, we didn't have to do anything to run on this architecture that a typical data scientist wouldn't do. In particular, we didn't write any CUDA code, as we did in Chapter 6.

Another interesting fact about the job flow that we just developed is that we loaded the data into the **landing_area** folder and started to learn the DNN model from the data. The flow that we built didn't generate any data. In other words, this methodology would work for any DL project, not just for a deep learning for numerical applications (DL4NA) project. (DL4NA is DL that speeds up analytics by approximating their results using a DNN). So by using an MTC framework such as SAS Infrastructure for Risk Management, you can easily mix and match DL, DL4NA, and any SAS programming. This flexibility allows you to gradually include DL and DL4NA into existing analytics by progressively replacing tasks and job flows with versions that leverage DL. The same thing could be said about gradually incorporating analytics written in CAS.

In summary, with SAS Infrastructure for Risk Management, we can easily write analytics that run in a heterogeneous architecture, because SAS Infrastructure for Risk Management seamlessly integrates a variety of hardware devices (CPU and GPU for now, and others in the future) as well as a variety of software systems (SAS 9.4, CAS, and DL, to name only a few).

Collaboration with Federated Areas

So far, we have created artifacts such as SAS programs and scripting client scripts using SAS Studio. But the files that we created are on disk under our PFA, as you can see in Figure 8.19.

Figure 8.19: The Development Artifacts of the PFA

Filename	Size
▼ 📁 pfa	--
▼ client_scripts	--
rw_scoring_flow.sas	1.3 KB
simple_flow.sas	584 B
rw_training_flow.sas	1.1 KB
etl_training.sas	582 B
etl_testing.sas	603 B
prepare_data_flow.sas	711 B
▶ generated	--
▼ landing_area	--
▼ testing	--
function_testing.sas7bdat	4.9 MB
▼ training	--
full_function_training.sas7bdat	24.2 MB
▶ input_area	--
▶ doc	--
▶ client_packages	--
▶ config	--
▶ jobflow	--
▼ source	--
▼ sas	--
▼ nodes	--
rw_score.sas	2.2 KB
rw_regression.sas	1.8 KB
prepare_testing_data.sas	1.1 KB
prepare_training_data.sas	1.2 KB
▼ ucmacros	--
irm_wordcount.sas	627 B
irm_normalize_data.sas	1.0 KB
▶ smd	--
▶ lua	--

In addition to the inputs that we just mentioned, the PFA also contains output files, such as the definition of the job flows, stored in BPMN files, as you can see in Figure 8.20. Those BPMN files are the output of the job flow creation scripts such as `rw_training_flow.sas` and `rw_scoring_flow.sas`.

Figure 8.20: Generated Artifacts of the PFA

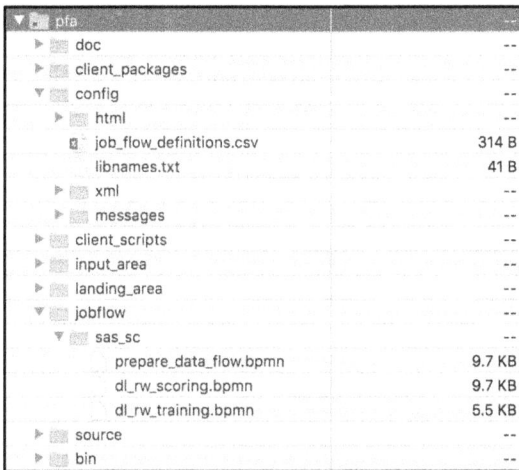

▼ 📁 pfa		--
▶ 📁 doc		--
▶ 📁 client_packages		--
▼ 📁 config		--
▶ 📁 html		--
📄 job_flow_definitions.csv		314 B
libnames.txt		41 B
▶ 📁 xml		--
▶ 📁 messages		--
▶ 📁 client_scripts		--
▶ 📁 input_area		--
▶ 📁 landing_area		--
▼ 📁 jobflow		--
▼ 📁 sas_sc		--
prepare_data_flow.bpmn		9.7 KB
dl_rw_scoring.bpmn		9.7 KB
dl_rw_training.bpmn		5.5 KB
▶ 📁 source		--
▶ 📁 bin		--

Let's assume that we are at a stopping point in the development of our analytics and that we want to share them with other SAS developers (on the same machine or on a completely different machine). One way to proceed is to create a compressed file of the complete PFA, as shown in Figure 8.21. In this case, we use a ZIP file, but as you can see in Figure 8.21, there are several alternatives.

Figure 8.21: ZIP File for the PFA

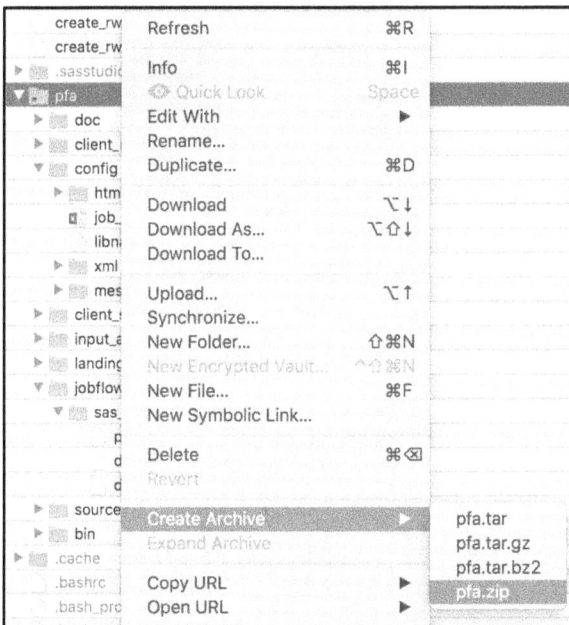

create_rw	Refresh	⌘R
create_rw		
▶ 📁 .sasstudi	Info	⌘I
▼ 📁 pfa	Quick Look	Space
▶ 📁 doc	Edit With	▶
▶ 📁 client_	Rename...	
▼ 📁 config	Duplicate...	⌘D
▶ 📁 htm		
📄 job_	Download	⌥↓
libn	Download As...	⌥⇧↓
▶ 📁 xml	Download To...	
▶ 📁 mes	Upload...	⌥↑
▶ 📁 client_	Synchronize...	
▶ 📁 input_a	New Folder...	⇧⌘N
▶ 📁 landing	New Encrypted Vault...	^⇧⌘N
▼ 📁 jobflow	New File...	⌘F
▼ 📁 sas_	New Symbolic Link...	
p		
d	Delete	⌘⌫
d	Revert	
▶ 📁 source	**Create Archive**	▶ pfa.tar
▶ 📁 bin	Expand Archive	pfa.tar.gz
▶ 📁 .cache		pfa.tar.bz2
.bashrc	Copy URL	▶ **pfa.zip**
.bash_pro	Open URL	▶

Now that we have a ZIP file, we can easily use its content to populate the PFA of another user. For example, let's assume that we have a user identified as DL4NA2. That user will also have a PFA that he or she can merge or replace with the content of the ZIP file that we just created. (See Appendix A to set up a new user if necessary at this point.) The exact mechanics of merging the ZIP file with DL4NA2's PFA varies from OS to OS.

After that copy or merge of the content of the ZIP file, things should look like Figure 8.22. DL4NA2 has a PFA with the content of DL4NA's PFA. The fact that DL4NA2's PFA looks like DL4NA's is already a good sign, but to convince ourselves that everything is indeed OK, let's run the script and see whether we get a successful flow execution.

Figure 8.22: DL4NA2 Is a New User

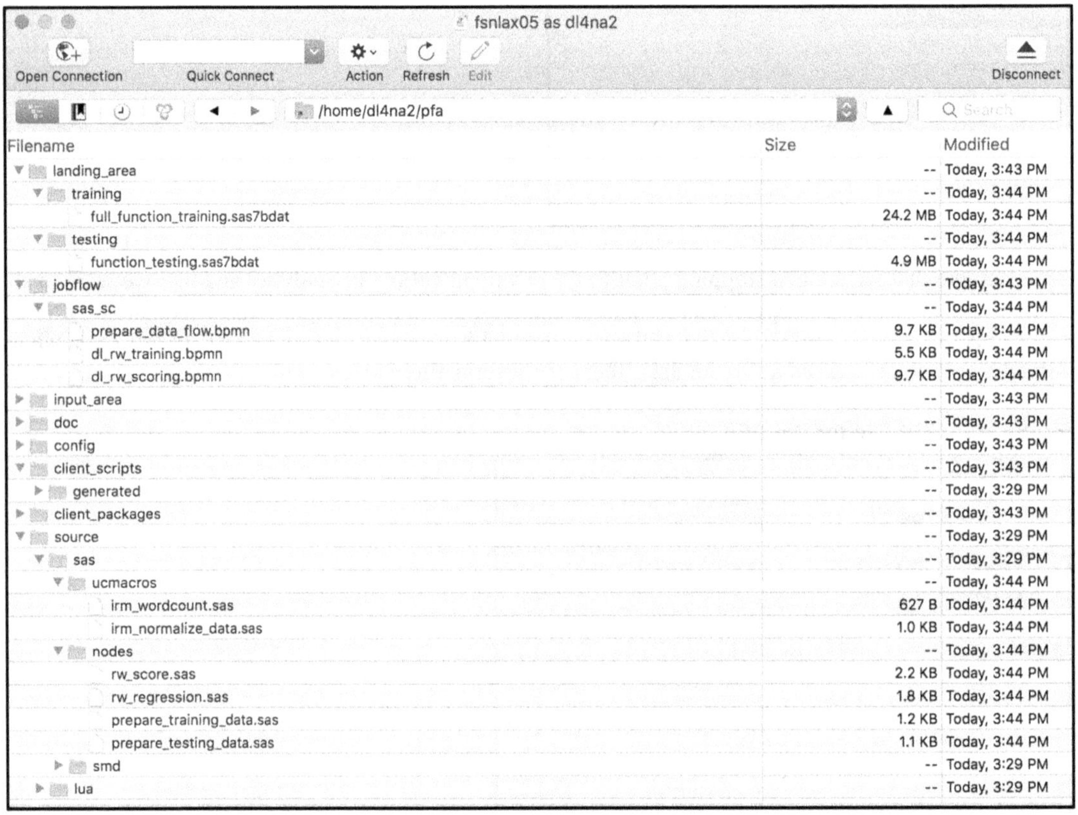

When we log in to SAS Studio, the interface should look like Figure 8.23. If we then run `rw_scoring_flow.sas` and go to the SAS Infrastructure for Risk Management GUI, we should see the same interface as Figure 8.24 (with tooltips and documentation), indicating that we successfully shared a PFA of one user with another user.

Note that if you run into file permission issues, it is probably because of the permissions associated with the files that you transferred using the content of the ZIP file. There are two main caveats to keep in mind.

First, in SAS Studio, everything runs as the logged-in user. In this case, that would be DL4NA2. So make sure that DL4NA2 can write to his or her own PFA.

Second, when running in SAS Infrastructure for Risk Management, everything runs as `sassrv` (or its equivalent server user). So make sure that `sassrv` can read DL4NA2's PFA (`sassrv` shouldn't have to write anything to the PFA unless you use live ETL).

Figure 8.23: A New User in SAS Studio

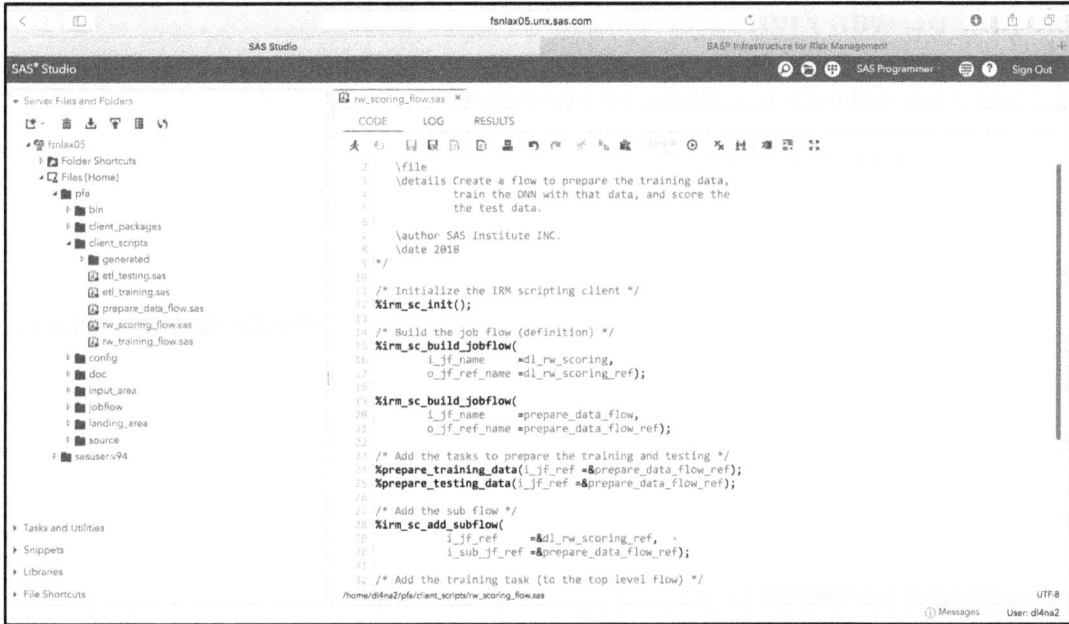

Figure 8.24: A New User in SAS Infrastructure for Risk Management

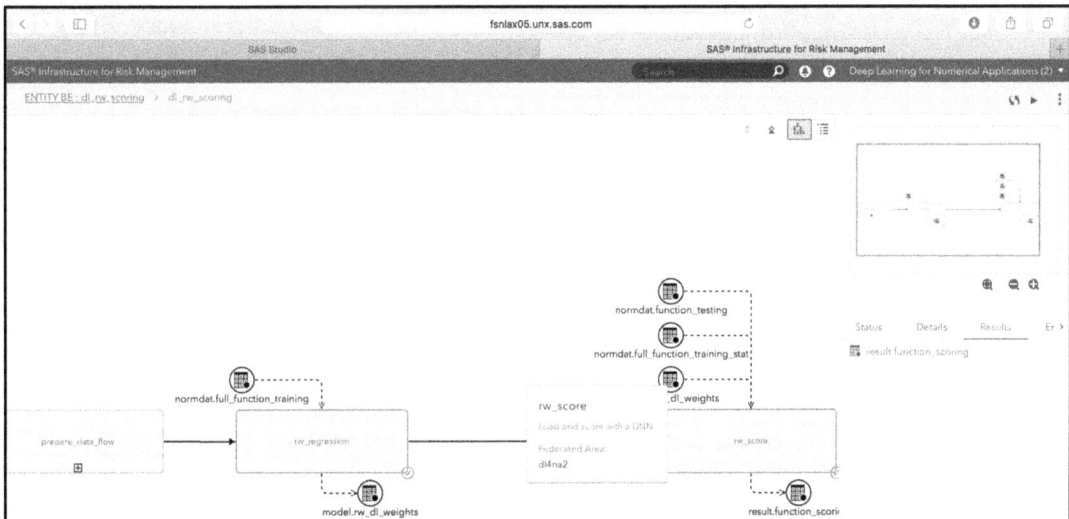

In addition to collaborating by exchanging the content (partial or complete) of PFAs, one can also collaborate using a source code control system such as GIT or CVS. Simply check files in and out and merge the PFA development artifacts that we discussed in this chapter.

Now let's focus on what happens when we are done with the PFA and are ready to deploy it to other users who are not necessarily SAS developers—users who want to run flows with different inputs and work with the outputs.

Deploying DL with Federated Areas

The content of the `pfa.zip` file that we used to share development artifacts between two SAS developers is usually simply referred to as "content," so that's what we'll do for the remainder of this chapter. This content can also be deployed on a more permanent basis. In this section, we go through the easy steps of deploying new content on a server.

If you examine the content of the ZIP file, you see something like Figure 8.25 (shown here on OSX).

Figure 8.25: ZIP File Content

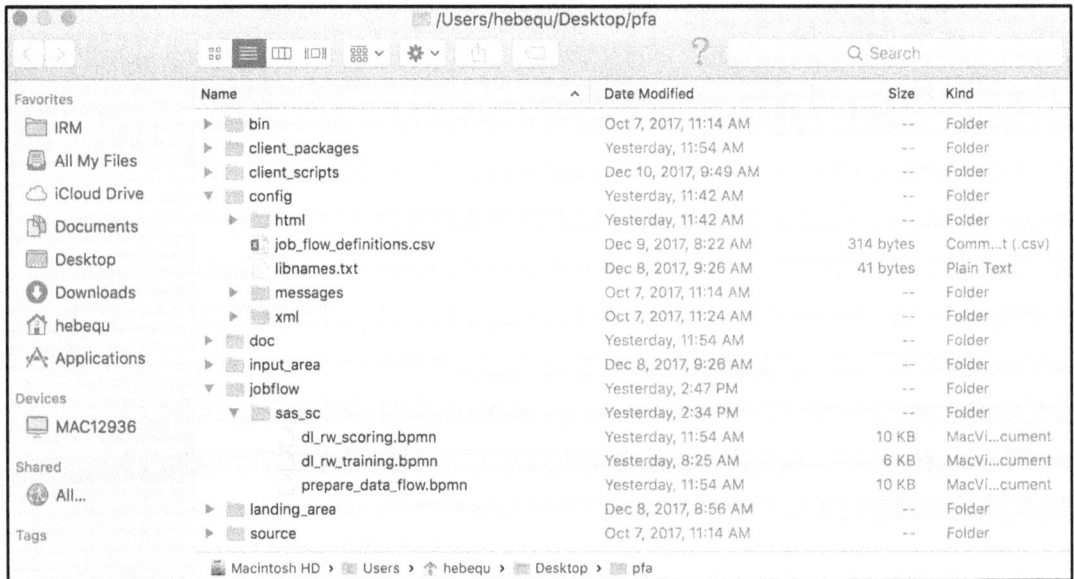

The folders that are open in Figure 8.25 contain the list of job flows and the configuration files that go with them. Note the **sas_sc** folder under **jobflow**. The **sas_sc** folder is called the category of the job flow, and it shows up in the SAS Infrastructure for Risk Management GUI in a drop-down menu that we will examine later. The name **sas_sc** is something that you might want to change to display something more appropriate than a "SAS Scripting Client" job flow category. In our example, we would like to change it to **random_walk**.

After exploding the ZIP file as you see in Figure 8.25, simply change the name of the folder to **random_walk**. In addition to the name of the folder on disk, you also need to change the name of the folder in the `job_flow_definition.csv` configuration file. To do that, modify the original content:

```
sas_sc,prepare_data_flow,solo
sas_sc,dl_rw_training,solo
sas_sc,dl_rw_scoring,solo
```

Change that content to this version using your favorite text editor:

```
random_walk,prepare_data_flow,solo
random_walk,dl_rw_training,solo
random_walk,dl_rw_scoring,solo
```

The `solo` keyword in this file can be ignored for our purpose; it merely indicates that the flow is for a single (solo) legal entity (for example, a company). The other possible value is `group`, to indicate that the flow is for a group of legal entities. More configuration parameters can be found in the SAS Infrastructure for Risk Management documentation.

Using this methodology, you can add as many folders (or job flow categories) as you want, but always make sure that the folders and the `job_flow_definition.csv` configuration file agree.

One subtle fact about flows is that they don't necessarily start at the **landing_area**: they might only include tasks that take their inputs from the outputs of other tasks. As you probably have realized as you're reading this, instances of those flows that don't start in the **landing_area** cannot run successfully (since they would be missing inputs). Having those flow definitions in the list of available flow definitions in GUI screens like Figure 8.27 doesn't make a lot of sense, so simply don't include them in the `job_flow_definition.csv` configuration file.

In addition to changing the job flow category names, you might want to change the display of some other artifacts in SAS Infrastructure for Risk Management. This is easy to accomplish with a properties file, **config/messages/jobflow.properties**:

```
# This file contains messages for development artifacts

# Job flow names
prepare_data_flow.bpmn = Prepare Data Flow
dl_rw_training.bpmn = Distance from the Origin in Random Walks (training)
dl_rw_scoring.bpmn = Distance from the Origin in Random Walks (scoring)

# Task names
prepare_training_data = Prepare Training Data
prepare_testing_data = Prepare Testing Data
rw_regression = Training
rw_score = Scoring
```

The `jobflow.properties` file can have multiple versions for different locales. For example, `jobflow_fr_FR.properties` could contain the translation of messages in French. SAS Infrastructure for Risk Management automatically picks up the correct file based on the locale of your browser. In the preceding example, we put labels only on job flow names and task names, but you can also put different labels on inputs and outputs. If you label the inputs and outputs, the tooltips in the GUI show the filename, so you can still easily make the correspondence between the GUI and the code or the SAS log.

Now that we have modified our content, we can repackage it into a new ZIP file, `random_walk.zip`, so that the name of the package reflects its content.

Installing the random walk FA entails the following steps:

1. Install the bits on the target machine; we use `/local/install/random_walk/` in this example.
2. Add the FA to SAS Management Console by navigating to **Application Management ▶ Configuration Manager ▶ SAS Application Infrastructure ▶ IRM Mid-Tier Server** (see Figure 8.26). You simply need a name for the FA and its path.
3. Restart SASServer8 where SAS Infrastructure for Risk Management is installed: `<SAS-Config>/Lev1/Web/WebAppServer/SASServer8_1/bin/tcruntime-ctl.sh restart`. Restarting SASServer8 is no longer required in IRM 3.5 (see IRM documentation for more details).

Figure 8.26: Adding the random_walk FA

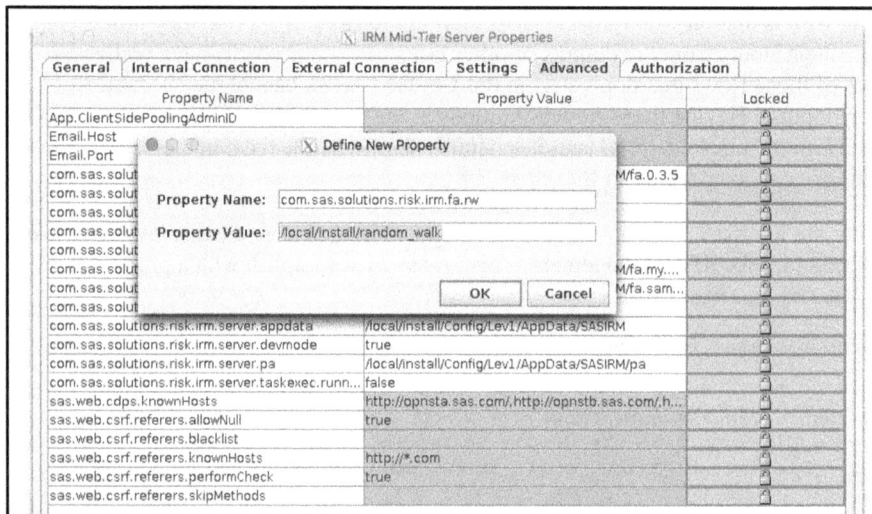

If you navigate in your browser to the SAS Infrastructure for Risk Management GUI and click the New Job Flow Instance icon (⬆), you should see something akin to Figure 8.27. Let's create an instance of our scoring job flow using the GUI. After the instance is created, if you drill down to it, you will see the display shown in Figure 8.28. As you can see in Figure 8.27 and Figure 8.28, the flow and task names pick up the labels from the `jobflow.properties` file. At this point, we have successfully deployed an FA that we developed. Note that a deployed FA is typically treated as read-only, because existing job flow instances rely on a stable job flow definition.

Figure 8.27: Creating an Instance with Messages

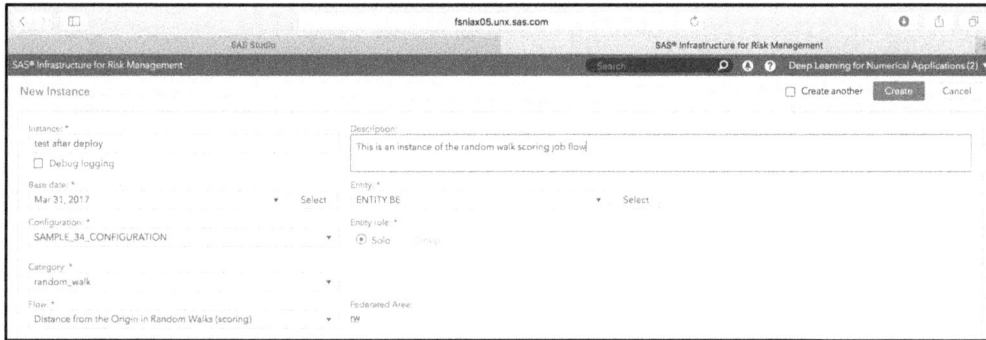

Figure 8.28: New Instance with Messages

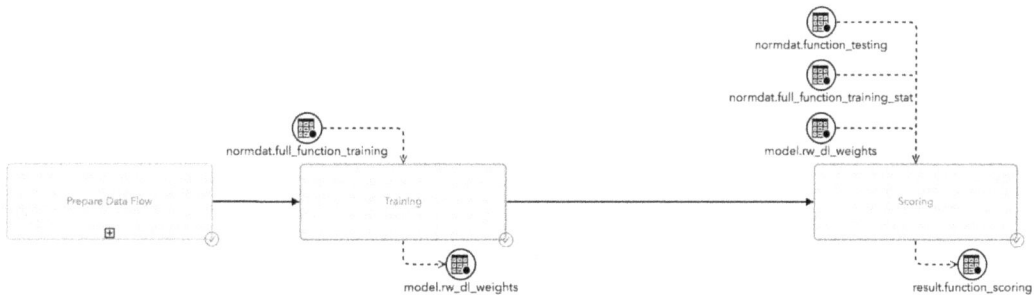

We should also point out that after an FA is deployed on a machine, SAS developers can still modify the content in their own PFA to supplement the deployed content. For example, you can create a new training flow to export a table with the iteration results for printing or plotting as we did in Chapter 2 and Chapter 3.

Conclusions

In this chapter, we took the skills that we learned so far in the book to collaborate on the development of deep learning analytics and to deploy our work so that it could be shared with others. In that endeavor, we relied heavily on the SAS Infrastructure for Risk Management many-task computing framework.

SAS Infrastructure for Risk Management enabled us to incrementally define machine learning workflows: first we prepared the data, then we defined the DL model, then we trained the model, and finally we scored our testing data using the DL model. Thanks to the data object pooling feature implemented in SAS Infrastructure for Risk Management, we could leverage the results of one incremental step during the development of the next. During scoring, for example, we didn't have to run through the preparation of the data or the training of the deep neural network.

To share the work between multiple SAS developers, we leveraged Federated Areas and Personal Federated Areas. When the development of our analytics was complete, we deployed the FA for full availability to all users, not just the SAS developers with whom we shared our PFA.

In the next and last chapter of this book, we will discuss additional applications of DL and future areas of research.

Chapter 9: Conclusions

In this concluding book, we look back at our accomplishments and give some pointers for the evolution of deep learning for numerical applications (DL4NA).

Data-Driven Programming

In 2007, Jim Gray of Microsoft Research gave a speech in which he argued that we were entering a fourth paradigm in science (Hey et al. 2009).

His argument was that science and research have gone through several phases:

1. **Experimental Science**
   This phase started thousands of years ago. It was mostly characterized by the observation of natural phenomena.

2. **Theoretical Science**
   This phase started a few hundred years ago. It began when scientists like Sir Isaac Newton and James Clerk Maxwell formally defined the phenomena that they were observing.

3. **Computational Science**
   This new phase started a few decades ago and is characterized by the simulation of complex phenomena. Chapter 5 in this book was our transition to computational science.

4. **Data-Intensive Science**
   This phase is sometimes called the data-driven science phase. It is characterized by an overwhelming avalanche of data (also known as big data) and by machine learning (ML). Chapter 7 in this book was our transition to data-intensive science.

The different phases are inclusive. For example, phase two includes phase one, since Sir Isaac Newton kept observing natural phenomena while he was writing *The Principia*.

An early adopter of the fourth paradigm of science, or data-intensive science, is astronomy. As astronomer Andrew Connolly, who focuses his research on understanding the evolution of our universe, eloquently put it in his TED talk: "Astronomy is constantly being transformed by this capacity to collect data" (Connolly 2014). Astronomy is in fact doubling its data every year, and the data stream is not slowing down. It is accelerating at a significant pace with projects such as the Large Synoptic Survey Telescope coming online (https://www.lsst.org/). In fact, an astronomer today is more likely to be online than looking at the stars with a telescope. A typical astronomer's workflow involves setting up observations, running them on a telescope in the mountains or in a desert or in space, and then downloading the results for analysis.

As another example of a scientific discipline being transformed by big data and ML, consider genomics. Genomics is a subfield of molecular biology that focuses on the characterization and quantification of genes. There is a lot of information in genomics: with an alphabet of 4 letters (A, T, C, G), it takes 3 billion letters to fully describe a single human being. In his TED talk, Riccardo Sabatini (a data scientist, not a geneticist) shows the impressive 262,000 pages of information that the genome of a single human being represents (Sabatini 2016). With this amount of information, the hope is that ML will help scientists unravel the significance of each letter to support personalized medicine.

But astronomy and genetics are certainly not the only examples of disciplines that have entered the era of big data in science.

Every year scientists and researchers gather in a conference called Super Computing, or SC, to exchange their views, solutions, and problems in computational science. The conference is typically identified by the year: the last conference before this book was written was called SC17. It took place in Denver, Colorado. At SC17, there were no fewer than 22 presentations and keynotes that had something to do with ML in general and deep learning (DL) in particular. There were actually many more presentations about DL, because it is often the motivation for heterogeneous architectures (more on this later). This is quite remarkable if you consider that the year before, at SC16 in Salt Lake City, there was a grand total of 2 DL presentations. In other words, it appears that we are strongly moving into the fourth paradigm of science.

But that is not all. The abundance of data and DL are also fueling hardware investments. For example, at SC17, NVIDIA announced the SATURN V supercomputer with 5,280 GPU V100 chips for 660 PFLOPS in DL performance. At 5,120 CUDA cores for each V100, that's 27 million CUDA cores!

These examples from the world of science can also be seen in the economy at large. The Internet of Things (IoT), for example, is an emerging new industry that exists only through the data that it can collect: without sensors and data, there would be no IoT. There are many more examples of data-intensive industries that regularly appear in the news, from self-driving cars to automatic translation to predictions of shoppers' behavior.

As you know, our focus in this book has not been to learn from the data as it is done in the examples that we just mentioned. In fact, we can summarize the motivation for this book with one word: speed.

The Quest for Speed

In this section, we examine the impetus behind this book and the path forward.

From Tasks to GPUs

In all the examples that we worked with, we looked at the speed of our analytics in multiple incarnations.

First, we ran single-threaded analytics. That gave us some performance numbers that we didn't quite like, so we graduated to parallelism with multi-threaded and multi-process executions. We saw orders of magnitude of performance improvements by going parallel. That speed increase came at a cost: complexity of development. We saw that by following the principles of task-based development and by using a many-task computing (MTC) framework such as SAS Infrastructure for Risk Management, we could tame that complexity and be productive by focusing our attention on our problems rather than the mechanics of multi-threading and multi-processing.

Organizing our code in tasks shed some light on the fact that not all tasks are created equal.

Some tasks take advantage of inherent independence in the processing of the data. Those so-called embarrassingly parallel or perfectly parallel tasks can speed up their execution by simply slicing up the data into partitions. Once the data is partitioned, each partition can be assigned to a processor (a core), and the scalability of our analytics mostly depends on the size of our machines, or how many processors we have at our disposal.

Not all problems are perfectly parallel, and we saw that we could automatically parallelize interdependent tasks by simply defining their inputs and outputs. We saw that immutable inputs would give us the best parallelization of our code, although at the cost of disk space. Speed versus space is not exactly a new problem in computer science.

The organization of our code into tasks has its limits in terms of parallelization. One of the major limitations is the number of CPU cores, which as of this writing is in the hundreds for a single machine, not in the thousands or in the millions. The general-purpose graphics processing unit (GPGPU or simply GPU), with its thousands of cores, addresses this limitation. We saw that GPUs are great at running the same algorithm on different data, which is a good fit for embarrassingly parallel tasks. We also saw that GPU programming using the CUDA framework is not trivial. The complex synchronization of multiple threads of execution using CUDA consumes a great deal of development resources. This programming complexity implies that CUDA is not the ideal tool for the data scientist or for the statistician.

Training and Inference

Fortunately, we realized that we could use DL to train a deep neural network (DNN) to give us reasonable approximations of our analytics. Scoring with a DNN for this approximation, also called inference, gives us another order of magnitude of performance improvement for our analytics. DL4NA was born.

We didn't worry about the time it took for training, because we didn't need to perform it very often. Another thing we didn't worry too much about, but probably should have, is latency: How quickly could we get a response?

Consider the use of DL4NA for approving loans over the phone or using a website. Many regulations govern the approval of a loan. For example, approving loans must be fair. You don't want to explain to regulators that you approved the loan for Jane, but not for Sue, when they have almost identical financial risk profiles. Another very important consideration for banks is their overall risk profile (across all loans held by the bank). By approving this loan, does the bank expose itself to a higher risk? For example, does the bank have too many loans in one geographic area? The risk profile must be calculated across an entire portfolio of loans, not just for one loan at a time. So for a bank with thousands or even millions of loans, this process of calculating the risk profile takes minutes, if not hours. This is clearly not desirable when you're waiting in your browser for an approval. This problem is solved by DL4NA: you train a DNN to approximate your decision of approving or denying a loan application in a couple of seconds or less. Clearly, DL4NA makes the decision on a loan application fast, but it will probably not satisfy a regulator, since DL doesn't easily give out its secrets (as we discussed in Chapter 2). To meet regulatory requirements, the bank can run through its entire risk profile once at the end of the day, not every time a new loan application is considered.

In the preceding example, we must take into account that we need to service tens, hundreds, or even thousands of inference requests at the same time. In such an application, latency of all the requests matters a great deal. If you go back to our example in Chapter 7, you'll notice that during inference, we would load the weights and biases first in the CPU and then in the GPU (the load of the parameters to the GPU was handled automatically by dlScore). Once the parameters of the neural network were loaded, we could proceed with the inference. Doing this process for every loan application is far from desirable. The latency will be in seconds, if not tens of seconds. Furthermore, the architecture of the GPU with its memory hierarchy and its off-ship memory is far from ideal because of the speed associated with the different types of memory defined by CUDA. If we had a device that could store the deep neural network parameter once and for all, things would go faster.

In fact, there is such a device: the field-programmable gate array (FPGA).

FPGA

An FPGA is a semiconductor that is based on a matrix of configurable logic blocks connected via programmable interconnects. In plain English, this means that an FPGA can be reprogrammed after manufacturing once it is in your hands or in your data center. In other words, an FPGA is a "soft" chip: silicon that you can program with software. This means that you can program the FPGA with the parameters of your deep neural network. In practice, you do this with the tools provided by the FPGA manufacturer.

Figure 9.1 shows a taxonomy of the CPUs, GPUs, FPGAs, and application-specific integrated circuits (ASIC) in terms of performance and development productivity.

Figure 9.1: FPGAs, CPUs, GPUs, and ASICs

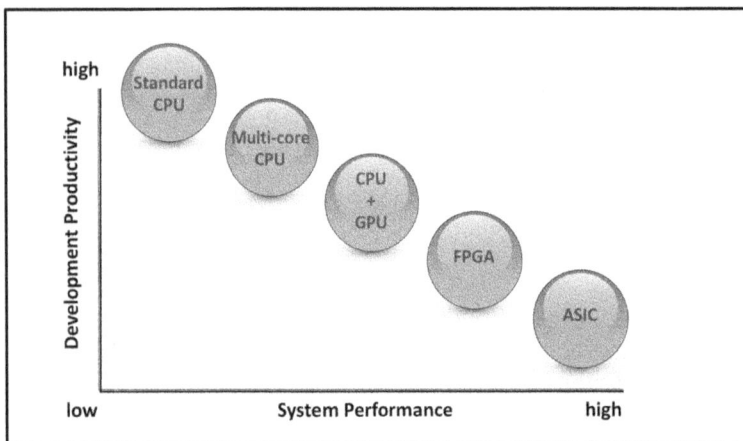

As you can see, FPGAs are harder to program than GPUs, but typically yield better system performance. Another way to look at this picture is to update our performance diagram that we introduced in Chapter 1 as you can see in Figure 9.2.

Figure 9.2: Performance of Analytics with FPGAs

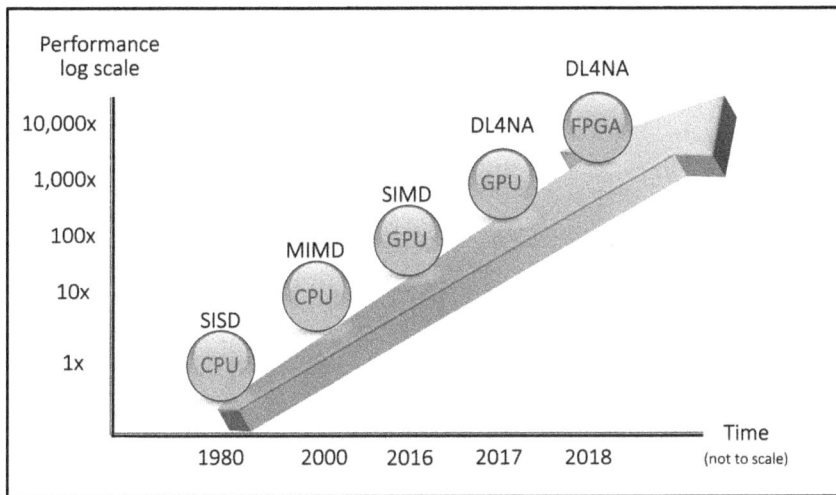

In addition to the latency problem, we have another problem with the GPU that perhaps you didn't notice: the floating-point operations per second (FLOPS) per watt. In the GPU runs that we did in Chapter 6, the output of `nvidia-smi` was something like this:

```
+-----------------------------------------------------------------------+
| NVIDIA-SMI 375.26                    Driver Version: 375.26           |
|-------------------------------+----------------------+----------------------+
| GPU  Name        Persistence-M| Bus-Id        Disp.A | Volatile Uncorr. ECC |
| Fan  Temp  Perf  Pwr:Usage/Cap|         Memory-Usage | GPU-Util  Compute M. |
|===============================+======================+======================|
|   0  Tesla K80            Off | 0000:05:00.0     Off |                    0 |
| N/A   62C    P0   149W / 149W |   345MiB / 11439MiB  |    100%       Default |
```

The operating temperature is 62° Celsius (143° Fahrenheit). That's not quite hot enough to cook an egg, but definitely hot to the touch. Imagine now that you have not one GPU card, but 8 or 16, and that they sustain this kind of load. As described in (Adhinarayanan, 2014), FPGAs typically have a much better FLOPS/watt ratio than GPUs.

This combination of CPUs and other devices such as the GPU or the FPGA is usually called a hybrid or heterogeneous architecture.

We should point out that for training, the GPUs are typically a better choice than the FPGAs because of their raw power.

Hybrid Architectures

In addition to the FPGA, Figure 9.1 also shows the characteristics of an ASIC: better system performance but even more difficulty for development. Better system performance usually means more FLOPS and a better FLOPS/watt ratio. In Figure 9.3 you can see the diagram from Figure 9.2 updated with ASICs. As of this writing, ASICs for DL are only in their infancy, and it is difficult to predict the level of performance improvements that they will provide. A 10x performance improvement is already attainable, so that's what we used in Figure 9.3.

Figure 9.3: Performance of Analytics with ASICs

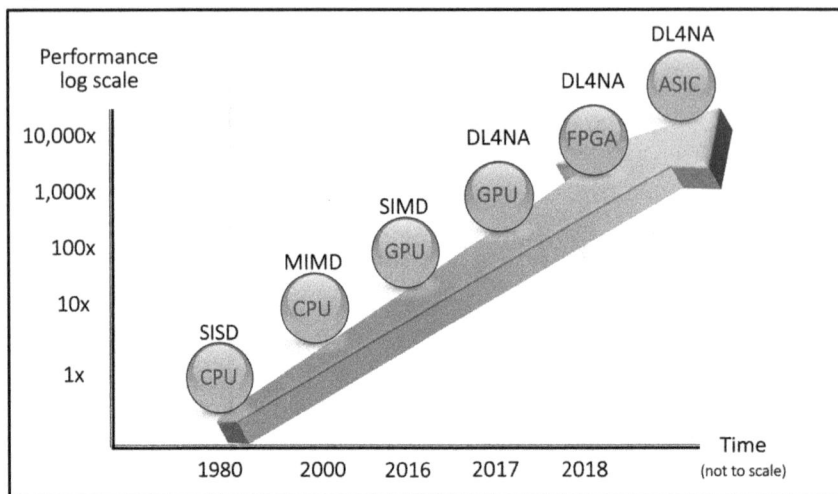

The natural conclusion of Figure 9.3 and this book is that if you want to deploy your analytics so that they run as fast as possible, you have to embrace hybrid architectures.

The good news is that with the techniques that we have developed in this book, you are well-armed to deploy your analytics on hybrid architectures. In addition, with the MTC paradigm, embracing hybrid architectures for the fastest analytics can be a smooth transition that doesn't invalidate your existing SAS infrastructure. We amply discussed this transition in Chapter 8.

Even though the emphasis of this book has been on the great performance gains with hybrid architectures, you can also benefit from a wider platform to deploy your analytics. With DL4NA, you can deploy *any* SAS program that you can approximate using a DNN to *any* of those devices. In particular, you could run an approximation of a SAS program in an IoT device, pushing your analytics to the edge of your network.

Appendix A: Development Environment Setup

In this appendix, we describe how to set up your development environment on LINUX and Windows.

LINUX

In this section, we explain how to add a new LINUX user to a SAS installation so that SAS Studio can be used to develop new many-task computing (MTC) flows with SAS Infrastructure for Risk Management.

We start by adding the dl4na user that will be in the same sas group as sassrv (or its equivalent). In a shell, enter the following commands:

```
[root@fsnlax05 hebequ]# useradd dl4na
[root@fsnlax05 hebequ]# passwd dl4na
Changing password for user dl4na.
New password:
Retype new password:
[root@fsnlax05 hebequ]# usermod -g sas dl4na
```

Now we add the dl4na user to SAS Management Console:

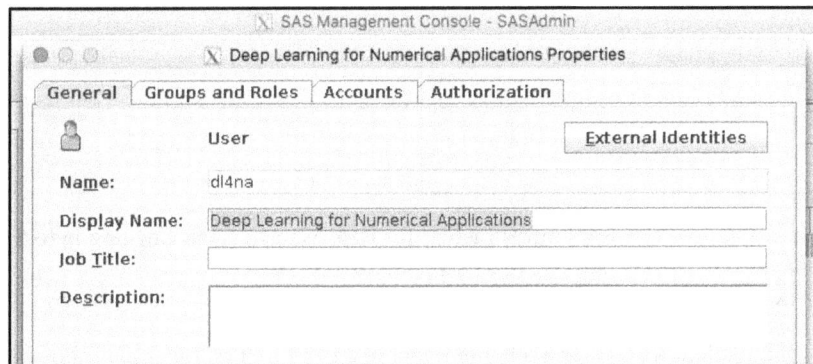

You also need to make sure that you enter dl4na's password (without that, SAS Studio won't be able to show inputs and outputs of SAS Infrastructure for Risk Management flows):

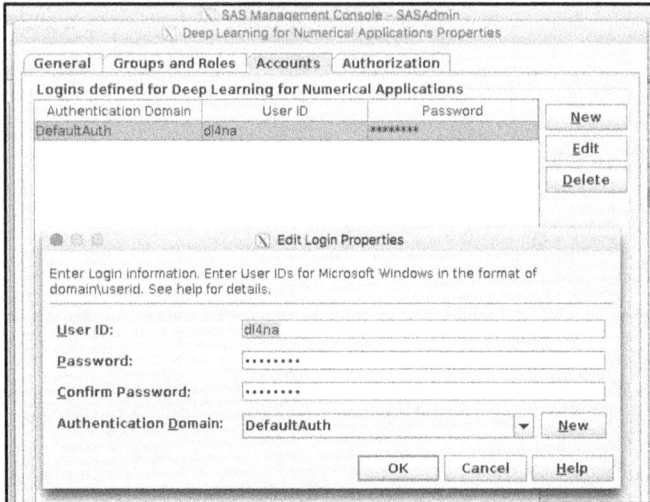

The dl4na user must also be added to the group/role IRM: Access All Entities (without that membership dl4na won't be able to log in to SAS Infrastructure for Risk Management):

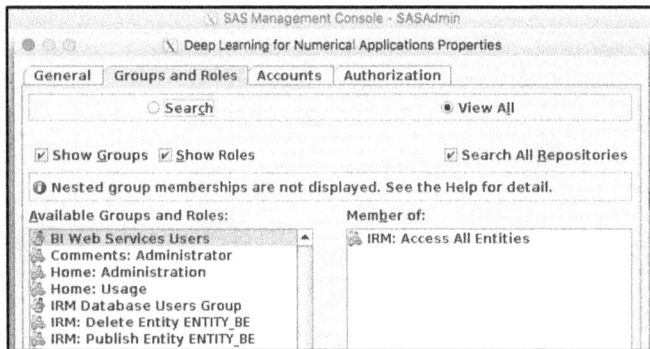

At this point, you should be able to log in to SAS Infrastructure for Risk Management. Logging in to SAS Infrastructure for Risk Management ensures that a PFA exists in **<SAS Config Folder>/Lev1/AppData/SASIRM/pa/fas/fa.dl4na**.

Now we want to create a symbolic link from dl4na's home folder to dl4na's PFA. Enter the following commands:

```
[dl4na@fsnlax05 ~]$ ls -ls
/local/install/Config/Lev1/AppData/SASIRM/pa/fas/fa.dl4na/
total 36
4 drwxrwxr-x 2 sassrv sas 4096 Sep 27 14:34 bin
4 drwxrwxr-x 2 sassrv sas 4096 Sep 27 14:34 client_packages
4 drwxrwxr-x 3 sassrv sas 4096 Sep 27 14:34 client_scripts
4 drwxrwxr-x 3 sassrv sas 4096 Sep 27 14:34 config
4 drwxrwxr-x 2 sassrv sas 4096 Sep 27 14:34 doc
4 drwxrwxr-x 2 sassrv sas 4096 Sep 27 14:34 input_area
4 drwxrwxr-x 2 sassrv sas 4096 Sep 27 14:34 jobflow
4 drwxrwxr-x 2 sassrv sas 4096 Sep 27 14:34 landing_area
4 drwxrwxr-x 7 sassrv sas 4096 Sep 27 14:34 source
```

```
[dl4na@fsnlax05 ~]$ ln -s
/local/install/Config/Lev1/AppData/SASIRM/pa/fas/fa.dl4na/ pfa
[dl4na@fsnlax05 ~]$ ls -ls
total 0
0 lrwxrwxrwx 1 dl4na dl4na 58 Sep 27 14:36 pfa ->
/local/install/Config/Lev1/AppData/SASIRM/pa/fas/fa.dl4na/
0 drwxrwxr-x 2 dl4na dl4na  6 Sep 27 14:34 sasuser.v94
```

In order to generate the documentation, we must install Doxygen: http://www.stack.nl/~dimitri/doxygen/download.html. Follow the download and install instructions. When you're done, you will have the `doxygen` executable on your machine, typically at `/usr/bin/doxygen`.

Our last step is to configure some advanced properties in SAS Infrastructure for Risk Management. Go to the Advanced IRM Mid-Tier Server Properties window and define the following properties:

- `com.sas.solutions.risk.irm.server.devmode` `true`
- `com.sas.solutions.risk.irm.sc.doxygen.path` `/usr/bin/doxygen`

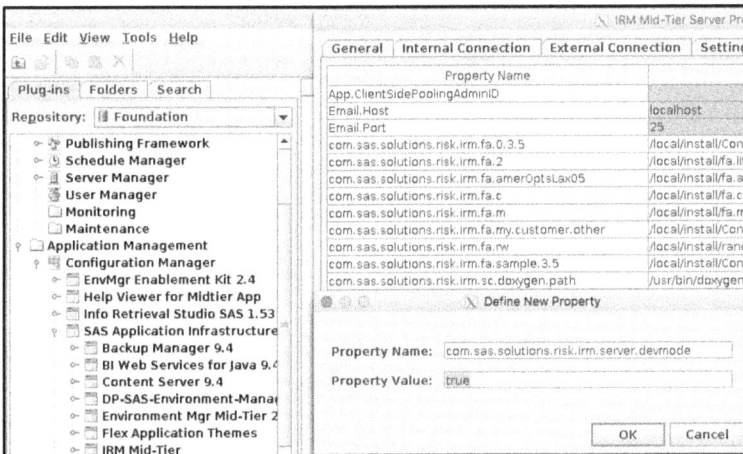

Now restart SAS Server **8**.

That's it! You should now be able to create SAS Infrastructure for Risk Management tasks and flows directly from SAS Studio after you log in as `dl4na`.

Windows

In this section, we explain how to add a new Windows user to a SAS installation so that SAS Studio can be used to develop new many-task computing (MTC) flows with SAS Infrastructure for Risk Management.

We start by adding the `dl4na` user. In Control Panel, select **User Accounts** and then **Edit local users and groups**. Then select **New user** from the pop-up menu. You can create a local or a domain account, but make sure that the OS name and the SAS Management Console name agree (the SAS Management Console account name must include the domain, but the user name doesn't have to). When you have created the Windows user, be sure to give the user the privilege to launch a batch job (otherwise, you won't be able to use SAS Studio with that user). To do so, in Control Panel, select **Local Security Policy ▶ Local Policies ▶ User Rights Assignment**, and select the user rights as shown in the following display:

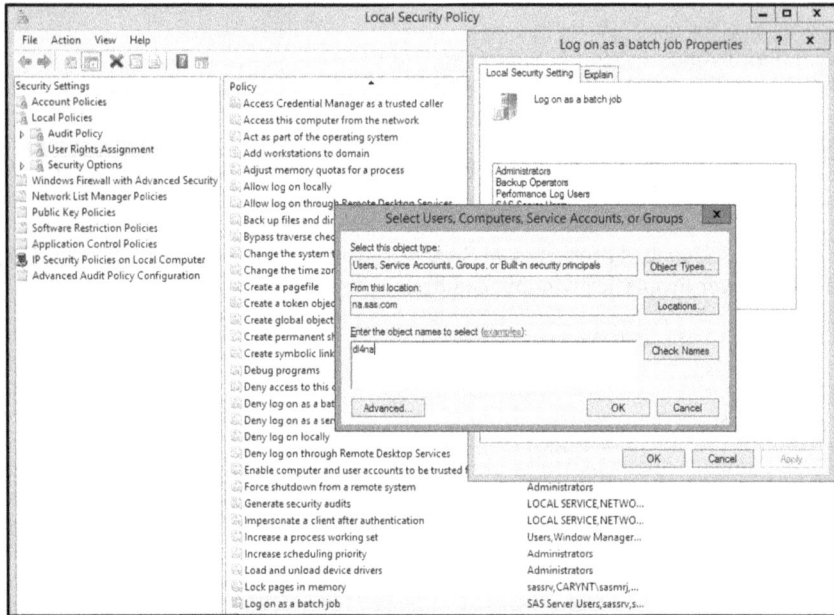

Now we add the `dl4na` user to SAS Management Console:

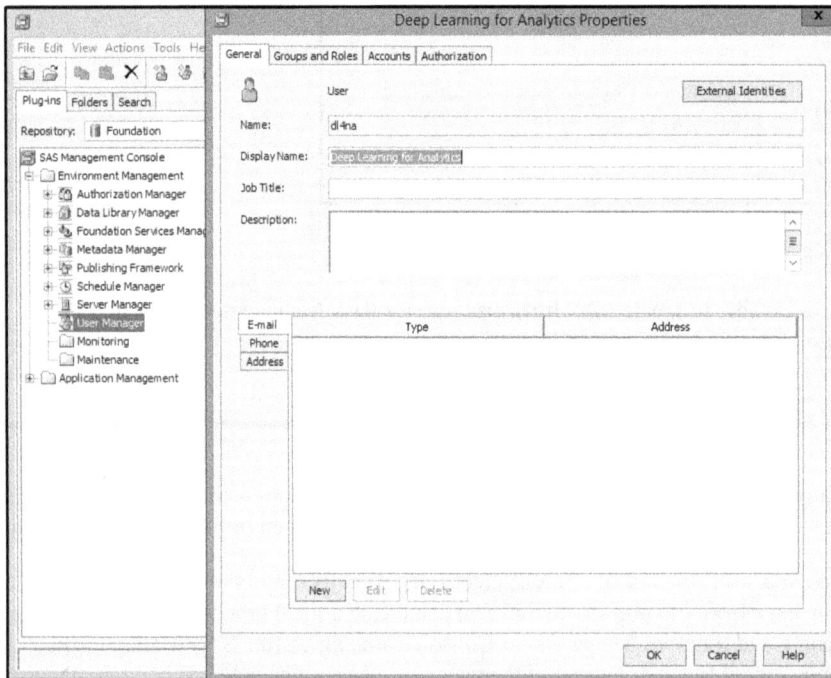

You also need to make sure that you enter dl4na's password (without that, SAS Studio won't be able to show inputs and outputs of SAS Infrastructure for Risk Management flows):

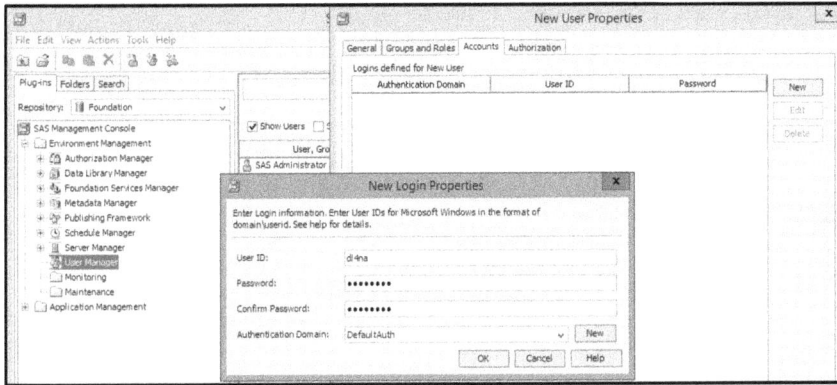

The dl4na user must also be added to the group/role IRM: Access All Entities (without that membership, dl4na won't be able to log in to SAS Infrastructure for Risk Management):

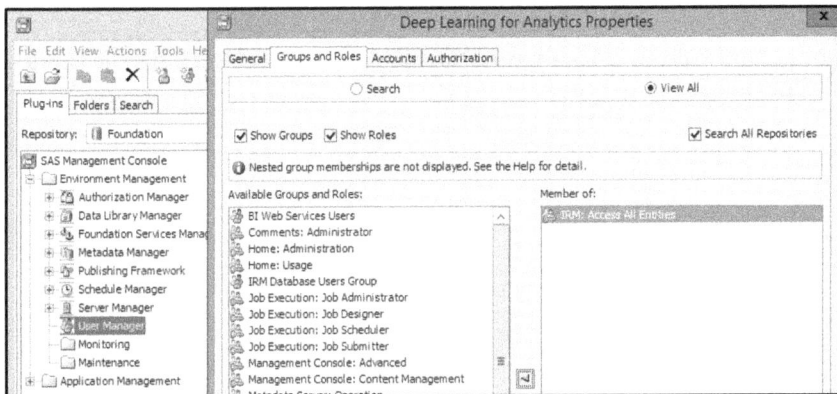

At this point, after restarting SAS Server 8, you should be able to log in to SAS Infrastructure for Risk Management. Logging in to SAS Infrastructure for Risk Management ensures that a PFA exists in `<SAS Config Folder>\Lev1\AppData\SASIRM\pa\fas\fa.dl4na`.

Now we want to create a symbolic link from dl4na's home folder to dl4na's PFA. Assuming that the current user is either an Administrator or dl4na, open a command prompt and enter the following command (shown in **bold**). This command changes the path to match your installation, if necessary:

```
C:\>mklink /D C:\Users\dl4na\Documents\pfa
C:\SAS\Config\Lev1\AppData\SASIRM\pa\fas\fa.dl4na
symbolic link created for C:\Users\dl4na\Documents\pfa <<===>>
C:\SAS\Config\Lev1\AppData\SASIRM\pa\fas\fa.dl4na
```

In order to generate the documentation, we must install Doxygen: http://www.stack.nl/~dimitri/doxygen/download.html. Follow the download and install instructions. When you're done, you will have the `doxygen.exe` file on your machine, typically at `C:\Program Files\doxygen\bin\doxygen.exe`.

Our last step is to configure some advanced properties in SAS Infrastructure for Risk Management. Go to the Advanced IRM Mid-Tier Server Properties window and define the following properties:

- `com.sas.solutions.risk.irm.server.devmode`
  `true`
- `com.sas.solutions.risk.irm.sc.doxygen.path`
  `C:\Program Files\doxygen\bin\doxygen.exe`

Now restart SAS Server 8.

That's it! You should now be able to create SAS Infrastructure for Risk Management tasks and flows directly from SAS Studio after you log in as `dl4na`.

References

Adhinarayanan, V., T. Koehn, K. Kepa, W.-C. Feng and P. Athanas. 2014. On the performance and energy efficiency of FPGAs and GPUs for polyphase channelization. International Conference on Reconfigurable Computing and FPGAs, December 8–10, 2014, Cancun, Mexico.

Arnow, B. J. 1994. On Laplace's extension of the Buffon needle problem. *The College Mathematics Journal* 25(1):40–43.

Badger, L. 1994. Lazzarini's lucky approximation of π. *Mathematics Magazine* 67(2):83–91.

Bequet, H. and H. Chen. Accelerate your SAS Programs with GPUs. Paper SASSD706-2017. Proceedings of the SAS Global Forum 2017 Conference. Available at: http://support.sas.com/resources/papers/proceedings17/SASSD0706-2017.pdf (accessed 18 June 2018).

Bressloff, P.C. 2014. *Stochastic Processes in Cell Biology*. New York: Springer.

Brin, S., L. Page. 1998. The anatomy of a large-scale hypertextual Web search engine. *Computer Networks and ISDN Systems* 30:107–117.

Buffon, G. 1733. Histoire de l'Académie Royale des Sciences, pp. 43–45. France: Académie des sciences.

Buffon, G. 1777. Essais d'arithmétique morale. *Histoire naturelle, générale et particulière*, Supplément 4:46–123.

Chen, H. and H. Bequet. 2017. Large portfolio variable annuity valuation powered by GPUs and deep learning. Cary, NC: SAS Institute, Inc.

Connolly, A. 2014. What's the next window into our universe? Available at https://www.youtube.com/watch?v=cI9vSC1eQ18 (accessed 1 June 2018).

Cybenko, G. 1989. Approximation by superpositions of a sigmoidal function. *Mathematics of Control, Signals, and Systems* 2:303–314.

Deng, J., W. Dong, R. Socher, L.-J. Li, K. Li, and L. Fei-Fei. 2009. Imagenet: A large-scale hierarchical image database. IEEE Conference on Computer Vision and Pattern Recognition, 2009. Available at: http://www.image-net.org/papers/imagenet_cvpr09.pdf (accessed 18 June 2018).

Elman, J. L., E. A. Bates, M. H. Johnson, A. Karmiloff-Smith, D. Parisi and K. Plunkett. 1996. Preface. In: *Rethinking Innateness: A Connectionist Perspective on Development* (Neural Network Modeling and Connectionism). Cambridge, MA: MIT Press.

Fisher, R. A. 1925. *Statistical Methods for Research Workers*. Edinburgh, London: Oliver and Boyd.

Fisher, R. A. 1936. The use of multiple measurements in taxonomic problems. *Annals of Eugenics* 7(2):179–188.

Flynn, M. J. 1972. Some computer organizations and their effectiveness. *IEEE Transactions on Computers* C–21(9):948–960.

Fowler, M. 2003. *Patterns of Enterprise Application Architecture*. Boston: Addison-Wesley.

Galton, F. 1886. Regression towards mediocrity in hereditary stature. *The Journal of the Anthropological Institute of Great Britain and Ireland* 15:246–263.

Goodfellow, I., Y. Bengio, and A. Courville. (2016). *Deep Learning*. Cambridge, MA: MIT Press.

Hastie, T., R. Tibshirani, and J. Friedman. 2009. *The Elements of Statistical Learning: Data Mining, Inference, and Prediction*, 2nd edition. New York: Springer.

Hey, T., S. Tansley and K. Tolle. 2009. *The Fourth Paradigm: Data-Intensive Scientific Discovery*. Redmond, WA: Microsoft Research.

Hinton, G. E., S. Osindero, and Y.-W. Teh. 2006. A fast learning algorithm for deep belief nets. *Neural Computation* 18:1527–1554.

Hinton, G. E., N. Srivastava, A. Krizhevsky, I. Sutskever, and R. R. Salakhutdinov. 2012. Improving neural networks by preventing co-adaptation of feature detectors. Available at: https://arxiv.org/abs/1207.0580 (accessed 1 June 2018).

Hopfield J.J. 1982. Neural networks and physical systems with emergent collective computational abilities. *Proceedings of the National Academy of Sciences of the United States of America* 79:2554–2558.

Hornik, K. 1991. Approximation capabilities of multilayer feedforward networks. *Neural Networks* 4:251–257.

Howarth, R. J. 2017. *Dictionary of Mathematical Geosciences: With Historical Notes*. New York: Springer.

Hubel, D. H., and T. N. Wiesel. 1959. Receptive fields of single neurones in the cat's striate cortex. *The Journal of Physiology* 148:574–591.

Izquierdo, J. 2017. Using IRM to price american style put options via CPU, GPU and neural network. Cary, NC: SAS Institute, Inc.

James, W. 1892. *Psychology, Briefer Course*. London: MacMillan & Company.

Karim, A., and S. Zhou. 2015. X-TREPAN: A multi class regression and adapted extraction of comprehensible decision tree in artificial neural networks. Available at: https://arxiv.org/ftp/arxiv/papers/1508/1508.07551.pdf (accessed 1 June 2018).

Kingma, D. P., and J. L. Ba. 2015. Adam: a method for stochastic optimization. Paper presented at International Conference on Learning Representations, May 7–9, 2015, San Diego, CA.

Knight, W. 2017. The dark secret at the heart of AI. *MIT Technology Review* Available at: https://www.technologyreview.com/s/604087/the-dark-secret-at-the-heart-of-ai/ (accessed 1 June 2018).

Komarov, N., and P. Winkler. 2013. Capturing the drunk robber on a graph. arXiv preprint arXiv:1305.4559.

Lazzarini, M. 1901. Un'applicazione del calcolo della probabilità alla ricerca sperimentale di un valore approssimato di π. *Periodico di Matematica per l'insegnamento secondario* 4: 140–143.

LeCun, Y. 2004. Paragraph on "Steep Learning Curves and other erroneous metaphors". Available at: http://yann.lecun.com/ex/fun/ (accessed 28 October 2017).

Lee, E.A. 2006. The problem with threads. Technical Report No. UCB/EECS-2006-1. Berkeley, CA: The Department of Electrical Engineering and Computer Sciences, University of California at Berkeley. Available at: https://www2.eecs.berkeley.edu/Pubs/TechRpts/2006/EECS-2006-1.pdf (accessed 8 June 2018).

Legendre, A.M. 1805. Nouvelles *Méthodes pour la Détermination des Orbites des Comètes*. Paris: Firmin Didot.

Love, R.M. 2010. The page cache and page writeback. In: *Linux Kernel Development*, 3rd edition, pp. 323–336. Upper Saddle River, NJ: Addison-Wesley.

Malkiel, B.G. 1973. *A Random Walk Down Wall Street.* New York: W. W. Norton & Company, Inc.

Metropolis, N., and S. Ulam. 1949. The Monte Carlo method. *Journal of the American Statistical Association* 44:335–341.

McCulloch, W. S. and W. Pitts. 1943. A logical calculus of the ideas immanent in nervous activity. *Bulletin of Mathematical Biophysics* 5:115–133.

Minsky, M., and S. Papert. 1969. *Perceptrons. An Introduction to Computational Geometry.* Cambridge, MA: MIT Press.

Nair, V. and G.E. Hinton. 2010. Rectified linear units improve restricted Boltzmann machines. In: *Proceedings of the 27th International Conference on Machine Learning, ICML 2010*, pp. 807–814. Madison, WI: Omnipress.

Ng, A. 2016. What artificial intelligence can and can't do right now. *Harvard Business Review* Available at: https://hbr.org/2016/11/what-artificial-intelligence-can-and-cant-do-right-now (accessed 1 June 2018).

NVIDIA 1999. NVIDIA Launches the World's First Graphics Processing Unit: GeForce 256. http://www.nvidia.com/object/IO_20020111_5424.html (accessed 18 June 2018).

NVIDIA 2017a. CUDA C Programming Guide. Available at: http://docs.nvidia.com/cuda/cuda-c-programming-guide/index.html (accessed 18 June 2018).

NVIDIA 2017b. NVIDIA Tesla V100 GPU Architecture. Available at: http://images.nvidia.com/content/volta-architecture/pdf/volta-architecture-whitepaper.pdf (accessed 18 June 2018).

Olazaran, O. 1996. A sociological study of the official history of the perceptrons controversy. *Social Studies of Science* 26: 611–659.

Orosi, G. 2015. A Simple derivation of risk-neutral probability in the binomial option pricing model. *International Journal of Mathematical Education in Science and Technology* 46:142–147.

Parker, M. 2017. Understanding peak floating-point performance claims. Available at: https://www.altera.com/content/dam/altera-www/global/en_US/pdfs/literature/wp/wp-01222-understanding-peak-floating-point-performance-claims.pdf (accessed 18 June 2018).

Pearson, K. 1905. The problem of the random walk. *Nature* 72:294.

Reinsel, D., J. Gantz, and J. Rydning. 2017. Data Age 2025: The Evolution of Data to Life-Critical. Available at https://www.seagate.com/www-content/our-story/trends/files/Seagate-WP-DataAge2025-March-2017.pdf (accessed 18 June 2018).

Rosenblatt, F. 1957. The Perceptron: A perceiving and recognizing automaton. Report 85-460-1. Buffalo, NY: Cornell Aeronautical Laboratory.

Ross, P.E. 2008. Why CPU frequency stalled. *IEEE Spectrum* 45: Issue 4, April 2008.

Rumelhart, D. E., G. E. Hinton, and R. J. Williams. 1986. Learning representations by back-propagating errors. *Nature* 323:533–536.

SAS Institute. 2018. *SAS 9.4 Functions and CALL Routines: Reference.* Available at: http://documentation.sas.com/?docsetId=lefunctionsref&docsetTarget=titlepage.htm&docsetVersion=9.4&locale=en (accessed 19 June 2018).

Sabatini, R. 2016. How to read the genome and build a human being. Available at: https://www.youtube.com/watch?v=s6rJLXq1Re0 (accessed 1 June 2018).

Satish, N., M. Harris, and M. Garland 2008. Designing efficient sorting algorithms for manycore GPUs. NVIDIA Technical Report NVR-2008-001.

Stigler, S.M. 1981. Gauss and the invention of least squares. *The Annals of Statistics* 9:465–474.

Tanaka, M., and O. Tatebe, Osamu. 2014. Disk cache-aware task scheduling for data-intensive and many-task workflow. 2014 IEEE International Conference on Cluster Computing (CLUSTER). Available at http://ieeexplore.ieee.org/document/6968774/ (accessed 20 June 2018).

Tzou, J., S. Xie, and T. Kolokolnikov. 2014. Drunken robber, tipsy cop: First passage times, mobile traps, and Hopf bifurcations. arXiv preprint, arXiv:1410.1391.

von Neumann, J. 1945. First Draft of a Report on the EDVAC. Philadelphia, PA: Moore School of Electrical Engineering, University of Pennsylvania. Available at https://sites.google.com/site/michaeldgodfrey/vonneumann/vnedvac.pdf (accessed 20 Jun 2018).

Weik, M.H. 1961. A Third Survey of Domestic Electronic Digital Computing Systems. Ballistic Research Laboratories. Report no. 1115.

Wicklin, Rick. 2012. The DO LOOP. Available at https://blogs.sas.com/content/iml/ (accessed 18 June 2018.

Zhang, C., H. Bequet, and J. Du. Disk Cache Aware Scheduling for Scientific Workflows. US Patent Application Serial No. 62/289,484. February 2016.

Index